Hot Science is a series exploring th[e latest ideas in science] and technology. With topics fro[m] dark matter to gene editing, these are books for popular science readers who like to go that little bit deeper ...

AVAILABLE NOW AND COMING SOON:

Destination Mars:
The Story of Our Quest to Conquer the Red Planet

Big Data:
How the Information Revolution
is Transforming Our Lives

Gravitational Waves:
How Einstein's Spacetime Ripples Reveal the Secrets
of the Universe

The Graphene Revolution:
The Weird Science of the Ultrathin

CERN and the Higgs Boson:
The Global Quest for the Building Blocks of Reality

Cosmic Impact:
Understanding the Threat to Earth from Asteroids
and Comets

Artificial Intelligence:
Modern Magic or Dangerous Future?

Astrobiology:
The Search for Life Elsewhere in the Universe

Hot Science series editor: Brian Clegg

NUCLEAR FUSION

The Race to Build
a Mini-Sun on Earth

SHARON ANN HOLGATE

ICON

Published in the UK and USA in 2022
by Icon Books Ltd, Omnibus Business Centre,
39–41 North Road, London N7 9DP
email: info@iconbooks.com
www.iconbooks.com

Sold in the UK, Europe and Asia
by Faber & Faber Ltd, Bloomsbury House,
74–77 Great Russell Street,
London WC1B 3DA or their agents

Distributed in the UK, Europe and Asia
by Grantham Book Services,
Trent Road, Grantham NG31 7XQ

Distributed in the USA
by Publishers Group West,
1700 Fourth Street, Berkeley, CA 94710

Distributed in Australia and New Zealand
by Allen & Unwin Pty Ltd,
PO Box 8500, 83 Alexander Street,
Crows Nest, NSW 2065

Distributed in South Africa
by Jonathan Ball, Office B4, The District,
41 Sir Lowry Road, Woodstock 7925

Distributed in India by Penguin Books India,
7th Floor, Infinity Tower – C, DLF Cyber City,
Gurgaon 122002, Haryana

Distributed in Canada by Publishers Group Canada,
76 Stafford Street, Unit 300,
Toronto, Ontario M6J 2S1

ISBN: 978-178578-922-9

Typeset in Iowan by Marie Doherty

Printed and bound in Great Britain
by Clays Ltd, Elcograf S.p.A.

For Micah,
in the hope that his generation
will benefit from fusion power

ABOUT THE AUTHOR

Sharon Ann Holgate is a freelance science writer and broadcaster with a doctorate in physics. She has written for newspapers and magazines, including *Science*, *New Scientist* and *The Times Higher Education Supplement*, and has presented on BBC Radio 4 and the BBC World Service. She was co-author of *The Way Science Works* (Dorling Kindersley, 2002), a children's popular science book shortlisted for the 2003 Junior Prize in the Aventis Prizes for Science Books, and recently completed the second edition of her undergraduate textbook *Understanding Solid State Physics* (CRC Press, 2021). She has also written the *Outside the Research Lab* textbook series (IOP Publishing in conjunction with Morgan & Claypool Publishers) and contributed to the popular science books *30-Second Quantum Theory* (Icon, 2014) and *30-Second Energy* (Ivy Press, 2018). In 2006, Sharon Ann won Young Professional Physicist of the Year for her work communicating physics. You can follow Sharon Ann on Instagram @everydaysciencethings

www.sharonannholgate.com

ACKNOWLEDGEMENTS

My thanks go firstly to series editor Brian Clegg, who I rather fittingly first discussed the possibility of this book with on the summer solstice, as well as the team at Icon Books, including Duncan Heath, James Lilford and Robert Sharman. I would also like to thank the numerous press officers, scientists and engineers who have given me many invaluable insights into the world of fusion energy and provided me with great support throughout this project. These include Laban Coblentz, Michael Loughlin and Mario Merola from ITER, Rob Buckingham, Chris Dorn, Nick Holloway and Nick Walkden from the UKAEA, Si-Woo Yoon from the KSTAR programme, Doug Parr from Greenpeace, Breanna Bishop from the LLNL, Daniel Alcazar and Alan Carr from LANL, Neal Singer and Dan Sinars from Sandia National Laboratories, Beate Kemnitz and Thomas Klinger from the Max Planck Institute for Plasma Physics, Mike Forrest, Randall Volberg from Type One Energy, Christos Stavrou of Fusion Reactors Ltd, Britta Weddeling and Jannik Reigl from Marvel Fusion, Tanner Horne from Horne Technologies,

Chris Ajemian, Scott Brennan and Derek Sutherland from CTFusion, Simon Redfern of Motive PR, Richard Dinan of Pulsar Fusion, Abbey Goodman of TAE Technologies, Jessie Barton from Helion Energy, Stephen Hasley-Mead from Tokamak Energy, Eric Lerner and Ivy Karamitsos from LPPFusion, Jan Kirchhoff and Warren McKenzie from HB11 Energy, Danielle Johnson from General Fusion and Chris J. Faranetta from NearStar Fusion.

Finally, I would like to thank friends and family who have helped in various ways with this book, including David Culpeck, Andrew Fisher, Paul Parsons and my mother Joan.

CONTENTS

INTRODUCTION
– HERE COMES THE SUN

From the earliest times, the Sun has held a fascination for us. Some early civilisations went as far as worshipping Sun gods – such as the Ancient Greek god Helios, who was believed to drive a chariot across the sky every day, while the Ancient Romans worshipped the Sun itself. Scientific study of the Sun also dates back centuries, with the Italian scholar Galileo Galilei one of the first to observe sunspots in 1610 using something invented just a couple of years before – a telescope. We are now in an era where there is a different type of focus on the Sun. For the past few decades, we have been striving to harness the same physical process that creates the Sun's energy – nuclear fusion. The aim is to provide our planet with a much-needed clean energy source and help meet global carbon reduction targets.

Of course, we are already harnessing solar energy for power generation via the familiar technology of solar cells. These provide a valuable renewable energy source, and scientists are currently working on developing a new generation of these useful devices. The intention is to create solar cells that generate more power, last longer, work in duller light and use light from a broader range of the Sun's spectrum. By contrast, in a fusion reactor we wouldn't be using the Sun's own energy to generate power. Instead, scientists and engineers are attempting to create what essentially amounts to a mini-Sun down here on Earth that operates at staggeringly high temperatures of around 150 million degrees Celsius (270 million degrees Fahrenheit).

This Sun-in-miniature would give out its own power via the nuclear fusion process. But we would not use its power directly. Instead, just as in current nuclear power stations, the heat generated by the nuclear reaction would be used to boil water, creating steam that drives turbines, which in turn produce electricity.

There are several features of fusion energy that make its use for electricity generation a very attractive prospect. Firstly, it is a clean energy source, emitting no greenhouse gases. Secondly, it could provide almost limitless quantities of power. It is also much safer than nuclear fission – which is the process that existing nuclear power stations use to generate electricity – and it creates no long-lived radioactive waste, which is currently a costly and problematic by-product of nuclear power generation. In addition, fusion power would enable countries to meet their own energy needs rather than importing either the raw materials to generate electricity or the power itself. In short, the development of fusion power plants would revolutionise global energy generation.

With the 2021 Intergovernmental Panel on Climate Change (IPCC) report stating that scientists are observing changes in the Earth's climate in every region of the planet, work towards new, clean energy sources seems more important, and more pressing, than ever. According to the report,

'Many of the changes observed in the climate are unprecedented in thousands, if not hundreds of thousands of years.' But it also says that 'strong and sustained reductions in emissions of carbon dioxide (CO_2) and other greenhouse gases would limit climate change'. So, there is still a chance to alter course, and many countries already have targets in place to cut the harmful emissions driving these unwanted changes to our planet. Fusion power plants could provide a powerful ally in reaching the ambitious emission reduction targets required.

Equally ambitious are the fusion experiments under way that may lead to this hoped-for new generation of power plants. The largest of these is the ITER ('the way' in Latin) experimental reactor being built in Cadarache in southern France. This has 35 nations pooling funding and scientific expertise – including the 27 member states of the European Union plus Switzerland and the UK, China, India, Japan, South Korea, Russia and the United States. ITER will never generate electricity, but it aims to demonstrate all the scientific and technical steps required to build commercial fusion energy power plants.

While ITER is by far the biggest of the multinational projects attempting to harness nuclear fusion for electricity generation, it is by no means the only contender in the race. Many countries have, or have had, their own research reactors. These include the Joint European Torus (JET) at the Culham Centre for Fusion Energy in the UK, which was first fired up in 1983, the now decommissioned Tokamak Fusion Test Reactor (TFTR) in the United States, which broke several records during its fifteen-year lifespan, the Chinese Fusion Engineering Test Reactor, which was powered up in December 2020, and the Korea Superconducting Tokamak

Advanced Research (KSTAR) project. KSTAR is a pilot device for ITER and, as we will see later when we look in more depth at these projects, set a new world record in November 2020 for one of the key stages in fusion reactor development.

Then, as we will also explore, there are the private companies fielding fusion contenders. Some of these have high-profile backers and collaborators, such as Canadian company General Fusion, which has funding from Amazon's Jeff Bezos, and TAE Technologies, which is in partnership with Google.

Not all of these projects are taking the same technological approach. As we will discover, there are different ways in which nuclear fusion reactions can be triggered and then contained, and multiple variants on the main approaches. No one is yet sure which approach will win out. Or, indeed, if a range of methods will be needed to help secure our future energy needs.

Along the way, the reactors look set to become spectacles in themselves. The sheer scale of ITER's reactor, for instance, with its 1 million components, weighing in at a combined total equivalent to the weight of three Eiffel Towers, is difficult to envisage. Meanwhile, General Fusion have employed award-winning architecture studio AL_A, whose previous commissions include the Victoria and Albert Museum Exhibition Road Quarter in London, to work on the design for their prototype power plant near Oxford in the UK.

Yet it's not just power stations that are on the cards. There is the potential for fusion-driven rockets to provide the transport for future interplanetary travel, including missions to Mars. And various technologies developed to help enable fusion energy studies to progress, such as the

This rendering shows how General Fusion's UK Fusion Demonstration Plant will look when completed.

AL_A for General Fusion

advanced robotic systems we will hear about in Chapter 7, have applications in other research or industrial settings. But the main goal of most fusion projects is to generate clean, sustainable power for our planet.

From the physics breakthroughs of the early 20th century that revolutionised our understanding of atoms, to the horrors of atomic weapons and the Cold War. And from the switch in the late 1950s to seeking peaceful applications of nuclear technology, to the latest experiments working towards fusion energy. This is the story so far of our quest to generate electricity from nuclear fusion.

WHAT IS FUSION?

1

The core of the matter

The challenge facing the teams trying to pave the way for nuclear fusion power plants is anything but small. To begin to comprehend its scale it is useful to first understand the fusion process. The best place to start with that is by looking at the atomic nucleus itself. This, by contrast, is most definitely small in nature.

It is amazing to think that just over 100 years ago, no one had even heard of atomic nuclei. Their existence was not suggested until May 1911, thanks to some experiments by students and colleagues of the New Zealand-born physicist Ernest Rutherford, carried out at the University of Manchester. Rutherford's analysis of their results provided a huge breakthrough, not least because it overturned in a stroke some of the groundbreaking earlier work by J.J. Thomson, the then head of the Cavendish Laboratory at the University of Cambridge (a position Rutherford would go on to fill in 1919, having previously studied there under Thomson from 1894 to 1897).

At the time of Rutherford's breakthrough explanation, the leading model for the atom was British physicist Thomson's so-called 'plum pudding' model. Physicists often use what are known as 'models' to explain complex processes. These models are descriptions of physical effects that help them visualise and study what is going on. In Thomson's model, every atom consisted of negatively charged particles, known as electrons, sat inside a spherical volume that was positively charged. The electrons could therefore be imagined to be a bit like the raisins dotted around inside a plum pudding with the main pudding mix representing the positively charged volume. Hence the model's nickname.

This had been a big leap in scientific understanding in itself. For centuries before, stretching right back to Democritus in Ancient Greece, scientists and philosophers had defined the atom as the smallest particle of matter that could exist. Thomson's experiments, in the closing years of the 19th century, had blown this idea out of the water. In 1897, he revealed the presence of electrons inside atoms while studying cathode rays. These then mysterious rays came from the negative electrode, known as the cathode, when a voltage was passed between it and the positive 'anode' electrode while both were sealed in a vacuum tube. At the time Thomson began these studies, some scientists thought that cathode rays were a form of radiation. Others believed them to be a stream of negatively charged particles. If the latter was the case, the particles would be deflected by both electric and magnetic fields, and you could work out what type of electric charge they had by observing the direction of the deflection.

Suspecting the cathode rays might indeed be composed of particles, Thomson applied an electric field to the tube via a second pair of electrodes. Sure enough, the cathode rays bent, and in a way that showed the 'rays' to be particles with a negative charge. Thanks to further experiments Thomson was able to determine the ratio between the charge and the mass of these negatively charged particles, which later became known as electrons.

This result kick-started a whole new area of scientific study – subatomic physics. But within a few years Thomson's student Rutherford was about to make another major step forward and show that the Thomson version of the atom was at best only part of the picture.

Having previously studied at Canterbury College, Christchurch, in his native New Zealand, Rutherford came to England to take up a scholarship at Cambridge University in 1895. Back home in New Zealand, he had been working on high-frequency magnetic fields. But after initially continuing this work at Cambridge, he switched to studying the effects that the then newly discovered X-rays had on air. In 1898, Rutherford moved country again, becoming a professor at McGill University in Canada. Here, he worked with Frederick Soddy on radioactivity.

The year after arriving at McGill, and before starting his collaboration with Soddy, Rutherford had shown that radioactive elements gave out two different types of emissions. He named these alpha rays and beta rays. By 1900, he had shown there was a third type of radiation emitted by radioactive substances – gamma rays.

It is difficult to overstate the impact this one man had on early 20th-century physics, and indeed on the story of fusion. Nuclear fusion experiments could simply not have come about without the fundamental physics discoveries made by Rutherford and his students and collaborators.

For instance, Rutherford's later work with Soddy revealed radioactive elements were changing into other elements – a process known as transmutation. While Soddy continued that work, Rutherford moved to looking at the alpha radiation emitted from radioactive elements.

In 1907, Rutherford had made another physical move, this time back to the UK to the University of Manchester. This was an exciting and fast-paced time for physics, and for Rutherford himself. Just a year after his arrival, he invented the Geiger counter with his colleague, German physicist Hans Geiger. Their initial version, the forerunner of the devices still used for detecting radioactivity today, was designed to detect the alpha particles they were studying. It consisted of a gas-filled tube threaded through with a wire along its longest axis. This wire had a high voltage running along it, and when alpha particles passed through the gas, they initiated a reaction that produced a pulse of current that could then be read on a meter.

With the help of their new device, Rutherford soon showed the alpha radiation (now known as alpha particles) was composed of helium atoms lacking their two electrons, which left them positively charged. It would take a few more experiments before it became clear just how significant this result was to be.

The 1909 experimental set-up that was to lead Rutherford to a breakthrough in understanding the structure of the atom was fairly simple. It consisted of a radon source, which is radioactive and gives out alpha particles, a lead screen with a small hole that let through a narrow beam of alpha particles and a thin piece of gold foil. These, along with a detection system, were housed within a cylindrical tube. Air was pumped out of this tube so that a vacuum was created inside, which allowed the alpha particles to travel further than they could have done through the air. The main component of the detection system was a glass screen coated with zinc sulphide, which emits a tiny flash of light whenever an alpha particle hits it. This detection screen was mounted at the end of a microscope that could be rotated through different angles.

Once the experiment was under way, the beam of alpha particles was directed towards the gold foil. The scientists expected to see the alpha particles either passing straight through the foil, or for their paths to be deflected slightly. The latter would occur thanks to the positively charged alpha particles interacting with the positive electric charge within the gold atoms of the foil in a similar way to the like poles of a magnet repelling each other. Any alpha particles shooting out from the foil at the correct angle to fall onto the zinc sulphide screen created tiny flashes of light as they hit the screen. So, by rotating the microscope with its attached detection screen around the apparatus, members of Rutherford's team were able to see what angles the alpha particles were being deflected by.

They observed that some of the alpha particles were indeed passing straight through the foil at the same angle they entered or were being slightly deflected by the positive charge within the gold atoms. The latter so-called 'scattering' of the particles was just as Rutherford had predicted. But what the team – which included Geiger and English–New Zealand physicist Ernest Marsden – had not bargained for was seeing flashes of light, indicating that a small number of the alpha particles had been bounced back in the direction they had come from. How could this be possible from the atom of Thomson with its positive charge spread around rather weakly throughout its volume?

The answer is it couldn't. Thomson's old 'plum pudding' model was now past its expiry date. These new observations could only be explained if atoms had a concentrated area of positive charge at their centres. The positively charged alpha particles would then be scattered off this region of positive charge like a ball bouncing off the edge of a pool table. This was exactly what was being seen in the experiment. So, in 1911, Rutherford proposed that every atom consisted of a positively charged nucleus surrounded by negatively charged electrons. Although this is a simplification, this is the basis of the model that is still taught to school pupils today. Rutherford would later state that the experimental result 'was almost as incredible as if you fired a 15-inch shell at a piece of tissue paper and it came back and hit you'.

As well as revealing the existence of the nucleus, this experiment also showed that it had to be small. Really small. This is because only a tiny percentage of the alpha particles were deflected by large angles, meaning that there was a low probability of colliding with a nucleus inside a gold atom. The vast majority of the alpha particles were not interacting with this positively charged nucleus and just sailed on through the gold foil. These results could only be possible if each nucleus took up a very small volume within each atom. In fact, we now know that the width of a nucleus is around 100,000 times smaller than that of the atom it sits within.

Quite literally in a flash, this experiment and the subsequent new description of the atom completely revolutionised physics. Fortunately for science, Rutherford was not done there.

Remaining neutral

In 1920, Rutherford predicted the existence of neutrons – subatomic particles with zero electric charge – which as we will see later have important implications for fusion reactors. By this point, Rutherford was head of the Cavendish Laboratory at Cambridge. In 1932, this was where his colleague, assistant director of research James Chadwick, discovered the neutron. British physicist Chadwick won a Nobel Prize 'for the discovery of the neutron' in 1935, an honour which Rutherford had himself received in 1908 'for his investigations into the disintegration of the elements, and the chemistry of radioactive substances'.

It was later figured out that nuclei consist of positively charged protons and electrically neutral neutrons. So, the simplest version of the currently accepted model of the atom, with its negatively charged electrons orbiting around a positively charged nucleus, is a product of these early 20th-century experiments and explanations.

As is often the case, there is an exception to the rule. In this instance, the model of the nucleus containing protons and neutrons holds for every type of atom except hydrogen. Hydrogen is different in that the nucleus of its most common form houses just the one proton with no neutrons to keep it company. If you look at the Periodic Table of the elements, you will see that hydrogen occupies the first spot in the table and is numbered one. You will also notice that the next element helium has the number two associated with it, lithium three and so on, rising in increments of one through the table. This number is the so-called atomic number. For a given element, the atomic number represents the number of protons in the nucleus of each of its constituent atoms.

Protons have a positive electric charge identical in size to the negative charge on electrons and, as we have just seen, neutrons are electrically neutral. Given that atoms isolated on their own are not electrically charged, this means that there are the same number of protons as there are electrons in an atom. Their charges simply cancel one another out. So, as the numbers of protons and electrons in an atom is equal, the atomic number can also represent the number of electrons in an atom.

Under certain conditions, the number of electrons in an atom can change, leaving the atom with an overall electric charge. This is a point we will return to shortly, as it is fundamental to creating nuclear fusion reactions. But before we look at how to trigger fusion, we first need to understand the fusion reaction itself. To do so, it is helpful to return to the times of Rutherford and his collaborators.

Delving deeper

While Rutherford was helping to change the face of physics at Manchester, back at McGill University, Frederick Soddy was continuing the work on transmutation that he had begun with Rutherford. This led him to make a discovery inextricably linked to neutrons – he found there were different variants of elements with differing numbers of neutrons in their nuclei. These variants are known as isotopes.

Isotopes are inherent to the fusion energy story because it is isotopes of hydrogen that will fuel most of the next generation of experimental fusion reactors. But it is not just hydrogen that has isotopes. Most of the elements in the Periodic Table have different isotopes. While every isotope of a given element has a different number of neutrons in their nuclei, they have the same number of protons. Taking hydrogen's three known isotopes as an example, we have already seen that hydrogen has one proton in its nucleus, so all three isotopes will have just the single proton. But while the most common version of hydrogen doesn't have any neutrons, the hydrogen isotope known as deuterium has one neutron, and the tritium isotope contains two neutrons.

The chemical properties of an atom are not changed by differences in the numbers of neutrons. The stability of the nucleus is, however, affected. This makes some isotopes, including tritium, radioactive.

Both protons and neutrons can be referred to as nucleons, and the number of particles in the nucleus, known as the nucleon number, gives a label for each isotope of an element. For example, carbon-12 has a total of twelve nucleons (six protons and six neutrons) in its nucleus, while carbon-14 has two more neutrons. Carbon-14 is a well-known and an extremely useful isotope. Not least because it is used for carbon dating of artefacts and trees, a process that was first proposed by American physical chemist Willard Libby in 1946, who went on to win the Nobel Prize for this development. Isotopes are used in a wide variety of settings. Nuclear medicine, for instance, uses a range of radioactive isotopes, including iodine-131 for treating thyroid cancer and technetium-99 during diagnostic scans, and the nuclear fuel used in fission reactors is a radioactive isotope of uranium known as uranium-235. But for the purposes of this story, we are mainly interested in deuterium and tritium, two of the most vital ingredients for achieving controlled fusion reactions.

Sometimes deuterium is referred to as 'heavy' hydrogen,

since the extra neutron in its nucleus means it weighs more than ordinary hydrogen. While the atomic number for all the isotopes of hydrogen is one, the mass number, which is the total number of protons and neutrons in the nucleus of an atom (also known as the 'nucleon number') for deuterium is two. Similarly, the mass number of tritium is three, accounting for the two neutrons and one proton in its nucleus.

What is particularly interesting is that if you add up the masses of the individual particles that make up a deuterium atom, you will find they weigh more than the atom itself. So, what is going on? At first sight this seems to make no sense, but it turns out that when neutrons and protons bind together to form a nucleus, they release energy. This comes about because they are in a more stable state, which has a lower energy, when they are joined together in a nucleus. Hence, they jettison the extra energy when they come together.

But how does this energy release relate to the mass difference? To understand this, we need to turn to arguably the most famous equation in the world. This is the mass–energy equation put forward in 1905 by German–Swiss–American theoretical physicist Albert Einstein:

$$E = mc^2$$

E stands for energy, m for mass and c for the speed of light. In the equation, c is squared, which leads to a very large number because the speed of light is approximately 3×10^8 metres per second, in other words 300 million metres per second.

It is because of this relationship between energy and mass that the nuclear reactions of both fission and fusion release energy. In nuclear fission – which was discovered in 1938, a year after Rutherford had died – heavy nuclei are split into lighter nuclei. By contrast, in the nuclear fusion process, light nuclei join together to form heavier nuclei. In both cases, because the speed of light squared is such a huge number, it is clear to see that there is an awful lot of energy released in the reaction.

Coming together

The basic concept of fusion is an idea we are all familiar with. Whether we are talking about glue fusing two pieces of paper together, fusion cuisine blending culinary traditions or fusion music merging different musical genres – the idea is the same. In all cases, things are combined together to make something more than the individual pieces.

In physics, the word fusion is used in a similar way since nuclear fusion involves two or more light nuclei fusing together to form a heavier nucleus, releasing both energy and subatomic particles in the process. As Einstein's equation reveals, we are talking about a really sizeable amount of energy coming from both fission and fusion reactions. For example, the amount of energy released per kilogram of the nuclear fission fuel uranium-235 is a startling 2–3 million times more than would be released from burning an equivalent amount of coal or oil. The energy yield from fusion fuel promises to be even greater, at around four times as much as that from an equivalent amount of fission fuel.

Because the two nuclei involved in the fusion process are both positively charged, just like the alpha particles being scattered by the positive charges of the gold atoms in Rutherford's Manchester experiment, the electric charges on the two nuclei act to force them apart. The best way to overcome this force pushing them away from one another is for the nuclei to gain lots of energy from moving around. They get this by being at very high temperatures so, to initiate fusion, the conditions must be hot. And by hot, we are talking about temperatures of around 150 million degrees Celsius (270 million degrees Fahrenheit). As we will discover later, reaching such temperatures is an engineering challenge in itself.

That challenge is made slightly easier by using atomic nuclei with low atomic numbers for fusion experiments. This is because the larger the atomic number of the nuclei involved in a fusion reaction, the more energy those nuclei need in order to fuse together so the hotter they need to be. It would simply not be practical trying to reach the temperatures required for the fusion of heavier nuclei. Even so, the light nuclei that are chosen for fusion must become so hot that they form a state of matter known as plasma.

Plasma is the fourth naturally occurring state of matter, the first three being solid, liquid and gas. (Other states of matter do exist, but they are laboratory-created.) Plasmas start life as gases and occur either when a gas is heated to an extremely high temperature, or when electric current is passed through a gas. In the former case, this is what we find in stars, including our Sun, and the latter is seen in the plasma balls used in classrooms and given as gifts. In both cases, the electrons are stripped away from their parent atoms, so the gas now contains equal numbers of separate electrons and positively charged atoms. Any electrically charged atom or molecule is known as an ion, and in this case the ion is positive because a negatively charged electron, which was keeping the atom neutral, has been removed. Within the plasmas artificially created for fusion experiments, the atoms in question are isotopes of hydrogen, so removal of an electron from these atoms leaves behind nothing but the positively charged nucleus. Because the temperature is so high, these nuclei moving around in the plasma have enough energy to overcome the forces pushing them apart so can fuse together.

Although a variety of different fusion reactions occur in stars, and can also be created in laboratories, there is one reaction which is the most efficient and therefore stands out as the primary candidate for generating fusion energy. This is fusion between a deuterium nucleus and a tritium nucleus. This reaction, illustrated below, results in the creation of a helium nucleus and a neutron, together with the release of a considerable amount of energy. The 'spare' neutron results because a helium nucleus contains two protons and two neutrons, but there are two protons and three neutrons coming together in the reaction.

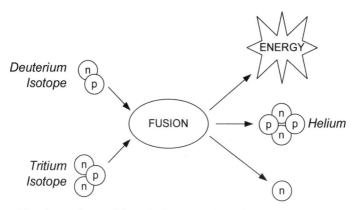

The deuterium–tritium fusion reaction gives out a helium nucleus, a neutron and a great deal of energy.

David Culpeck

It was fusion experiments through the second half of the 20th century, some of which we will explore in the next chapter, that confirmed this so-called D–T reaction (where D stands for deuterium and T for tritium) to be the fusion reaction that gives out the most energy at the lowest temperature. That being said, the 150 million degrees Celsius temperatures needed to initiate it are, incredibly, ten times hotter than the temperature required for the hydrogen reaction taking place in our Sun as we do not have the advantages of the immense gravitational forces compressing the fusion fuel together. (A point we will return to later in Chapter 5.) Nor do we have the Sun's huge physical size and density, which result in the nuclei not being able to escape colliding with each other many times and eventually undergoing fusion.

Star maker

Our Sun is not the only natural fusion reactor. It is fusion reactions that create the energy in all stars. Inside a star's core, elements are also being created via nuclear fusion. But not all stars are working the same way; there are several different types of fusion processes that can take place. These will differ depending on the age of the star, its size and composition.

The sorts of temperatures you will find in a young star like our Sun, of around 15 million degrees Celsius (27 million degrees Fahrenheit) at the core, are exactly what are needed for fusion to take place there. In all young stars there is hydrogen in the centre. Because of the enormous temperatures and pressures, this is in the form of plasma, which consists of hydrogen nuclei and separate electrons. The hydrogen nuclei hit into one another in the hot, dense centre of the star, and some of them undergo fusion reactions when they collide. They fuse together to form helium, a reaction which simultaneously releases vast amounts of energy. Our Sun has been creating helium and generating energy via this process for almost 5 billion years.

In fact, that is not the whole story as a peculiar quantum mechanical effect known as quantum tunnelling can help the nuclei overcome their electrostatic repulsion, and thereby lead to more fusion in stars than would result from high temperatures and pressures alone. One way of visualising quantum tunnelling is to first think about the forces pushing the nuclei apart as an energy barrier that must be overcome. If you then imagine this barrier as a small hill and a nucleus as a ball, if the hill was fairly low and you could give the ball a good push, it might be able to roll right over the top of the hill. But if the ball did not have enough energy to make it over the hill, it would simply roll back down again, remaining on the side where it started its journey.

In this scenario, the ball would be either one side of the hill or the other depending on how much energy it had from its motion. But atoms and any particles smaller than atoms do not behave like balls. The motion of balls is described by classical physics, whereas atoms and subatomic particles are governed by quantum physics. This means that thanks to the inherent weirdness of quantum mechanics – where probabilities reign supreme – there is a very small chance that a nucleus without enough energy to get over an energy barrier can in fact make it to the opposite side of the barrier to where it started out. It is as though the nucleus has tunnelled its way through the barrier. By some of the Sun's nuclei 'tunnelling through' their electrostatic repulsion from one another, additional fusion can take place.

The Sun is classed as a small- to medium-sized star. All stars begin their life as clumps of matter moving around in space. As they near one another, gravity causes them to clump together tightly, forming a star. There are two main forces acting on a star. There is an inwardly directed force of gravity created by the star's mass and an outward force from the energy generated in the fusion reaction happening in the star's core. Throughout most of a star's lifetime, this inward-facing force balances the outward-facing force.

Over time, the hydrogen in the centre of a star starts to run out, and at this point there are helium nuclei in the core colliding with one another. Eventually, hydrogen fusion ceases. Without the pressure from the hydrogen fusion, the core of the star then has to shrink in size in order to keep the forces acting on it in equilibrium. This shrinkage, which astronomers and cosmologists call gravitational contraction, increases the pressure and temperature of the core considerably. In fact, the temperature rises to around 100 million degrees Celsius (180 million degrees Fahrenheit), which combined with the pressure increase is enough to set off a new round of fusion reactions involving the helium nuclei.

But again, this cannot last for ever, as there is only a finite amount of helium available. Something has to give.

In smaller stars like our Sun, once the helium supply runs low things are nearing the end. The star develops instabilities, which lead to its outer layers becoming a shell of gas that expands further and further outwards, eventually dispersing into space. This leaves behind the star's core which shrinks and cools down becoming a white dwarf.

For larger stars, the core can shrink again and the

temperature become so high that further fusion reactions, involving heavier elements, can take place. In fact, elements along the Periodic Table right up to iron can be formed via this process repeating itself, depending on the size of the star. It ceases at iron because to fuse iron energy would need to be put into the star. And once energy needs to go in, the star can never be in equilibrium, so it begins to die. With all its nuclear fuel now exhausted, the core contracts until it eventually implodes and sets off a supernova explosion.

Reaching an understanding

These stellar processes first began to be revealed less than a decade after Rutherford was carrying out his experiments on alpha particles and had discovered the existence of nuclei. In 1916, the British astrophysicist Arthur Eddington began focusing his research on the interiors of stars, and ten years later, one of the key tenets of our understanding of stellar reactions – that the gravitational force acting inwards must be exactly balanced by the outwardly directed pressure forces from gas and radiation – appeared in his book *The Internal Constitution of the Stars*.

Eddington was also the first to suggest that it was radiation, not as previously believed convection, that transferred the heat energy from the centre of a star to the outer areas. By 1920, Eddington had suggested that it was the fusion of hydrogen into helium that was responsible for the incredible amount of energy coming from stars.

His model for fusion seemed to tally pretty well with the energy output coming from our Sun. But it couldn't account for the output seen from larger, hotter stars. The mystery was partially solved by German–American physicist Hans Bethe, who published an astrophysics paper in 1938 that provided more details on the fusion reactions creating elements in large stars. This corresponded with data on the energy being given out by large, hot stars. But Bethe's work did not explain the origin of the carbon-12 that was the catalyst for the cycle of fusion reactions he was proposing. It would be another nineteen years before that part of the puzzle was completed.

Carbon, with six protons, is produced in stars such as our Sun in the helium fusion part of their life cycle. Up until the 1950s, the prevailing idea was that three helium nuclei would fuse together into a carbon nucleus. However, British astronomer Fred Hoyle, whose studies in the 1950s encompassed the formation of elements in the universe, realised that this just didn't stack up. There was simply too much carbon in the universe for it to have been produced by this reaction that only happened infrequently. So, what was going on?

Hoyle began studying another mechanism that, rather than comprising just one fusion reaction, consisted of two stages instead. The first involved two helium nuclei fusing to form beryllium. This was then followed by a beryllium nucleus undergoing a fusion reaction with a helium nucleus to create the carbon. Hoyle, along with colleagues, eventually worked out the details of how this reaction was taking place. They published their results in a landmark 1957 paper, which described how the elements heavier than hydrogen and helium are created within stars.

It is a sobering thought that at the same time Hoyle was trying to get to grips with the particulars of stellar fusion, research was pressing ahead on using hydrogen fusion to create weapons. The United States tested its first hydrogen bomb on 1 November 1952, with the Soviets following a few months later. Fortunately, by the time Hoyle and his collaborators published their seminal results, scientists and engineers had already begun looking into the feasibility of recreating some of these stellar fusion processes for the peaceful use of electricity generation. They were soon to discover that releasing as much energy as possible in a very short time frame – an essential requirement for a bomb – is a much easier feat to achieve technologically than a controlled fusion reaction. The latter case involves much more complex systems that can contain, sustain and harness plasma in a fusion reactor.

While a variety of ideas were tried out initially, as we will hear in the next chapter, nowadays, there are two main experimental approaches for fusion energy.

Emulating nature

Whatever method is to be used, when it comes to creating a fusion power plant, the plan is to create what basically amounts to a mini-Sun here on Earth. In order to generate electricity from nuclear fusion, the reactor will first have to achieve a state of 'ignition'. This is the point at which the fusion reaction becomes self-sustaining and about four times more energy is released compared with nuclear fission. The heat created thanks to this reaction would then convert water into steam, which would be used to drive giant turbines that generate electricity.

No one is yet certain exactly what design of reactor might prove to be the winning formula for a fusion power station.

But although there are a range of potential designs, they can all be classified into one of two methods for achieving fusion. These two approaches are known as magnetic confinement and inertial confinement.

In the former, a plasma of fusion fuel is created and then ignited, creating the temperatures needed for fusion to occur. This super-hot plasma is held away from the reactor walls using incredibly strong magnetic fields. This is possible because the charged particles (ions and electrons) that make up the plasma create their own magnetic fields as they move around, which then interact with the applied magnetic field. The external magnets act to force the charged particles along magnetic field lines that effectively form a cage to confine the plasma.

By contrast, inertial confinement involves firing lasers or particle beams at a small pellet of hydrogen fuel. This compresses the pellet to such a high density that the hydrogen nuclei are pushed into one another and fusion occurs.

There is no getting away from the fact that the technology which led to this point was created on the back of atomic weapons research. But exactly how did something so destructive turn into a purely constructive project? To find out we need to look back to the final year of the Second World War.

WRITTEN IN THE STARS 2

Out of control

Horrifyingly, Cold War politics in the late 1940s and early 1950s led to the United States creating a new type of nuclear bomb first tested in November 1952 at Enewetak Atoll in the Pacific Ocean. Their new weapon had hundreds of times more explosive power than the atomic weapons they had dropped on Japan in 1945. This was the hydrogen bomb.

Thankfully, hydrogen bombs, also known as H-bombs or thermonuclear weapons, have never been used in warfare. But they remain the deterrent of choice for nuclear nations today.

Hydrogen bombs work in a different way to the atomic bombs that were used to such devastating effect in Hiroshima and Nagasaki at the end of the Second World War. The latter bombs, called Little Boy and Fat Man, were based on nuclear fission and were shockingly powerful.

When Little Boy exploded over Hiroshima on 6 August 1945, it released the equivalent amount of energy to

15,000 tons of TNT exploding. It flattened the city, and although estimates of casualties vary, it is thought Little Boy killed well over 100,000 people, while Fat Man ended the lives of at least 74,000 people in Nagasaki three days later. Fat Man was even more powerful than Little Boy, with an explosive force of over 21,000 tons of TNT (although the type of terrain at Nagasaki meant it caused less destruction than Little Boy). In stark contrast, a hydrogen bomb could unleash the equivalent of tens of millions of tons of TNT. Its explosive power dwarfs that of fission bombs. So, what is enabling such a massive energy release, and how does this relate to future power stations?

In answer to the first part of this question, it is thanks to fusion of hydrogen nuclei that hydrogen bombs release such a spectacular amount of energy. Inside the bomb, the hydrogen nuclei fuse together to form helium nuclei, releasing vast amounts of energy in the process.

There are two main types of hydrogen bomb. Two-stage fission–fusion bombs, and more powerful three-stage fission–fusion–fission bombs. The former type has a fission bomb in its centre which raises the temperature so high that fusion is triggered in the surrounding deuterium or lithium deuteride. In a three-stage weapon, the fusion reaction is used to explode a third fission stage.

Hydrogen is the perfect choice for the fusion stage in such bombs as it only has a relatively weak positive electric charge so there is less resistance to fusion (via repulsion of like charges) than there would be from heavier nuclei. In fact, it is deuterium and tritium, which as we saw in the last chapter are isotopes of hydrogen, that are used inside hydrogen bombs. The very same fuel that could one day be powering fusion power plants.

However, in the 1940s and 1950s, the main focus was on weaponry rather than power generation. Within the space of a few years, the physics and engineering developed in the US H-bomb project had successfully created an extremely effective destroyer of lives. Other nations, including the Soviet Union, were also developing their own hydrogen weapons. In short, the world was fast moving towards a very scary place. But at the same time, people began to wonder whether all that energy – both literally and metaphorically – could be put to a peaceful use that would help the human race rather than risk annihilating it.

Thanks to some welcome political developments, physicists and engineers were soon able to start putting the technological expertise they had gained during the H-bomb programmes to use on a whole new range of fusion energy experiments. This time they would be aiming to create a powerful new source of electricity generation that could benefit the whole planet.

Peaceful purposes

At 2.45pm on Tuesday 8 December 1953 in Geneva, Switzerland, Dwight D. Eisenhower, then President of the United States of America, began delivering a speech that would shape the world. His address, to the 470th Plenary Meeting of the United Nations General Assembly, is known as the 'Atoms for Peace' speech. It included a plan for the United States to 'encourage world-wide investigation into the most effective peacetime uses of fissionable material' as well as seeking to reduce the dangers posed by atomic weapons. This was game-changing, and fusion energy studies are just part of the peaceful research that it led to, although it would be a few more years before nations collaborated on their research.

Initially, fusion energy research was a top-secret activity because at that time anything that produced neutrons was considered a weapons source. Equally, much of the early technology and expertise was transferable to nuclear weapons programmes. Against this political backdrop, in 1957, Mike Forrest, one of the UK's fusion energy pioneers, applied for a job at Harwell.

This Oxfordshire site was then the home of the UK Atomic Energy Authority Research Establishment, and Forrest's application to work there had come about completely serendipitously. As Forrest was coming up for completing his sandwich course studying applied physics and electronics at university, Eric Pulsford, an electronic warfare specialist from Harwell, happened to be invited to give a talk to Forrest and his fellow final-year students. The soon-to-be graduate, who was on the lookout for interesting work opportunities, was so captivated by Pulsford's talk that he approached him afterwards to enquire how he could apply for a position at Harwell. Pulsford arranged for an application form to be sent, and in due course Forrest found himself going through the security checks for entering the top-secret research establishment when he arrived for interview.

Probing enquiry

Forrest – who had carried out two work placements at Rolls Royce Aero Division, surveying the temperatures inside the combustion chambers of a new type of jet engine, as well as a stint with the Central Electricity Authority, testing power station equipment – impressed the panel. He was offered a job, subject to security clearance, on the day.

This was a far cry from entering fusion research today, as Forrest had no idea what job he had been offered or even what research area he would be working in. Due to the secrecy surrounding the research at Harwell, the post was completely classified. This meant he did not discover until his first day at work that he had been assigned to Harwell's controlled thermonuclear fusion research team, which had started its research in the late 1940s.

'When they interviewed me for the job, the main thing they liked about me wasn't the physics I knew but that I was interested in high-quality lens design and other technical aspects of photography [optics]. They realised optical techniques were going to be one of their main diagnostics,' recalls Forrest.

This was because a lot of the early work on fusion was focused around understanding the behaviour of plasma, including its temperature. Because plasma is so hot, any temperature probes put into it would simply melt. But with no way of measuring plasma temperatures, our knowledge of plasmas would not be able to progress very far. A solution was clearly needed.

Whenever scientists are faced with the inability to measure something directly, they look instead for indirect methods of making measurements. One familiar example is using ultrasound to look at unborn babies in the womb, which gives medics a lot of useful information without them needing to physically handle the child. In this case, to investigate the temperatures within the incredibly hostile environment of a plasma, the solution was spectroscopy – which is where Forrest and his optics know-how came in.

There are a range of different types of spectroscopy, all of which, as the name suggests, involve looking at the spectrum of light (or other electromagnetic radiation) coming from atoms, particles or other types of matter under investigation. Most spectroscopic methods involve probing that matter with a beam of light or other radiation and looking at what gets bounced back. The laser-based technique Forrest began developing in a new laboratory in Culham, Oxfordshire, along with cosmic-ray expert David Evans from Bristol University and plasma physicist Alan DeSilva from the University of Maryland, was no exception.

We saw in Chapter 1 that plasmas start life as gases that have either been heated to very high temperatures or have had an electric current passed through them, which causes electrons to be stripped off from their parent atoms in the gas. The resulting plasma therefore consists of equal numbers of positively charged atoms (ions) and separate electrons. The hotter a plasma is, the faster the electrons in it will be moving. So, if you shine a laser into a plasma, the spectrum of the laser light scattered back off these electrons will change depending on the speed they are moving at. Analysing the light which is scattered back therefore enables you to measure the plasma temperature.

In fact, the scattered light undergoes Doppler broadening of its spectrum, similar to the change in sound frequency that we hear from the siren of an emergency vehicle as it races past us. (The Doppler effect is named after the Austrian physicist Christian Doppler (1803–53), who discovered in 1842 that the observed frequency of a wave of electromagnetic radiation or sound changes depending on the relative motion between the source of the wave and where it is being measured or experienced.) This type of scattering – of electromagnetic radiation from electrons, which are free to move about or are only very loosely associated with their parent atoms – is known as Thomson scattering, named after the creator of the plum pudding atomic model we met in the last chapter.

Within a couple of years of the head of their division suggesting the use of Thomson scattering with lasers, the three-man Culham team – Forrest, Evans and DeSilva assisted by two technicians and a selection of engineers – were developing sophisticated laser spectrometers. This was remarkable progress given that lasers had only just been invented and that they had first needed to create a bespoke laser capable of probing their plasmas. Once Forrest and his colleagues could burn a hole through a stack of several razor blades, they realised they had the necessary power.

Armed with their new laser, which had taken only six months to build, they were soon up and running and taking the temperature of their tricky patient. But there was more to come. Further information was buried in the scattered light they were analysing. This was revealed by some of Forrest's colleagues who were theoretical astrophysicists. They could model the light waves emitted and work through the equations of scattering theory to not only get readings of the temperature, but also determine the density of the plasma.

All these efforts were assisted in no small way by this experimental project being funded by a blank cheque. 'In those days, money was no object. We had total carte blanche to do what we liked,' recalls Forrest, adding that their world-first, accurate measurements of the temperature and density of the plasma were subsequently published in *Nature*.

This method of measurement is still used in fusion research today, so without question the efforts of Forrest and his colleagues changed the face of plasma physics internationally. Not least because, as we will hear shortly, this work led to an unlikely collaboration with Soviet physicists at the height of the Cold War.

But before we hear about Forrest's journey behind the Iron Curtain, it is worth understanding the stage experimental fusion reactors were at in the UK and the United States.

First steps

The UK Atomic Energy Authority (UKAEA) laboratory at Culham, built on the site of a former Royal Navy airfield a few miles south of Oxford, opened in 1965. This was a purpose-built centre for fusion research, and in its first eighteen years of operation, over thirty different experiments to test out various reactor design concepts had taken place.

The first fusion devices were so-called 'pinch' machines that used strong magnets to pinch, or compress, plasma with the aim of creating a fusion reaction. At Harwell, the approach being taken to fusion in the late 1950s was a 'Z-pinch' design. This consisted of a single coil of wire encased in a deuterium-gas-filled tube. This coil had a high voltage passed through it that created plasma from the gas, as well as a magnetic field that pinched the plasma away from the walls. The origins of this linear device had been weapons research, and the machine operated from 1957 to 1968 – initially under a cloak of secrecy.

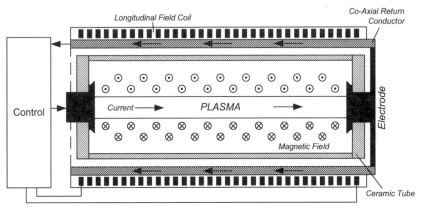

The Z-pinch concept.
David Culpeck

There was also a UK machine based on the 'plasma focus' approach. This design had a 5-centimetre (2-inch) diameter solid cylindrical conductor nested inside a 10-centimetre (4-inch) diameter, 25-centimetre (10-inch) long cylindrical electrode in a vessel filled with a low-pressure gas of hydrogen or deuterium. A high voltage of 40,000 volts was discharged across one end of the cylinders, which caused sparks between the outer cylindrical electrode and the inner conductor. These sparks created plasma from the gas. The resulting plasma was very dense, and more importantly for its original purpose in weapons research, neutrons were produced at the same rate as in a nuclear bomb. The plasma focus method had been independently invented by the Russians and the United States, and the UK version was based on the US design. Although the birthplace for the device had been in weapons programmes, for Forrest and his colleagues, it proved to be an extremely useful test bed for developing their pioneering laser plasma diagnostics techniques.

It is fair to say that there was a lot of optimism surrounding fusion work in the late 1950s, exemplified by the feeling that commercial power generation via fusion was not far away. There were a lot of different approaches to the problem, and the UK also constructed the Zero Energy Thermonuclear Assembly (ZETA). It was built at Harwell at a cost of £300,000 – around £5.1 million ($6.4 million) in today's money. ZETA, which was first started up in 1957, managed to reach temperatures of 5 million degrees Celsius (9 million degrees Fahrenheit). This temperature created plasma from deuterium gas inside a ring-shaped metal tube. The device quickly broke previous records for keeping plasma alive. Prior to ZETA, the only plasmas created in fusion machines were over within just a few microseconds. ZETA achieved results 1,000 times better, keeping plasmas going for more than a millisecond.

The plasma was 'confined' in ZETA by magnetic fields that pushed it away from touching the inner walls of the tube and melting them. Magnetic fields can be produced by magnets and by electric currents. The latter was an effect first discovered by the Danish physicist Hans Christian Ørsted (1777–1851). He observed that bringing a compass near to a wire carrying current caused the needle to move. As we will see in the next two chapters, this effect is exploited in several types of fusion reactor in which the magnetic fields produced by an electric current are used to help shape and confine plasma.

In the case of the ZETA machine, the strongest magnetic field came from the electric current running through it rather than from ZETA's magnets. Meanwhile, a contemporary Soviet fusion reactor design known as a tokamak – which was getting much better results than any Western machines – had the opposite idea of using stronger fields from external magnets and less field from the currents to contain its plasmas.

This design of fusion reactor had been conceived in the 1950s by the Soviet physicist Andrei Sakharov, who subsequently collaborated with another Soviet physicist Igor Yevgenyevich Tamm on the design. The word '*tokamak*' comes from an acronym for the transliteration of a Russian expression for a toroidal (doughnut-shaped) chamber with magnetic coils, which as we will see in Chapter 3 is essentially what the design consists of.

The UK scientists working on ZETA were lagging behind their Soviet counterparts thanks to a couple of blunders. Firstly, the UK team incorrectly thought that the magnetic field due to the current in the plasma was the important thing – hence their different magnetic field arrangement. (It later turned out to be external magnets combined with an internal field that was crucial.) They also thought that passing an electric current through the plasma would be enough to heat it. In fact, you need external heating as well.

'We were all a bit naïve in those days. We didn't realise how complex a plasma is,' explains Forrest, adding that it was difficult to advance the understanding of the physics until their laser technique was developed because prior to that, 'we couldn't even define the plasma properly'.

One positive aspect relating to these blunders by the UK team was that they were kept behind closed doors, unlike an unfortunate episode that occurred shortly after ZETA began operating.

Hitting the headlines

To begin with, ZETA had used hydrogen as its fuel. But during the August of its first operating year, the Harwell researchers switched to using the hydrogen isotope deuterium, and they were increasing the current going into the machine. Then, in the evening of 30 August 1957, everything changed.

While the plasma was being held at 5 million degrees Celsius, suddenly, thousands upon thousands of neutrons began flowing out of the plasma. Since this quantity of neutrons could indicate fusion had been achieved, surely this was the breakthrough they had been looking for?

Some certainly thought so, and leaks from this supposedly top-secret project began being reported by the press. In January 1958, Harwell finally held a press conference to announce their results to the world. At the helm of the ZETA programme was the Atomic Energy Authority Research Establishment director, the Nobel Prize-winning British physicist Sir John Cockcroft (1897–1967). During the press conference, he sounded a cautious note, warning that they had not definitely proved that the vast numbers of neutrons they were seeing were coming from fusion. But Cockcroft's admission that he was '90 per cent certain' they were a result of a fusion reaction sent the media into a frenzy. Headlines screamed out the achievement: 'A Sun of Our Own and It's Made in Britain!' proclaimed the *Daily Sketch* on Saturday 25 January 1958, while the *Daily Mail* ran with 'Unlimited Fuel for Millions of Years'.

Some international figures in fusion research were sceptical, and they were to be proved right. By the spring of 1958, the Harwell researchers had to release a correction to their January announcement. The neutrons they had detected had turned out to be the result of collisions between high-energy electrons and ions in the plasma. They were not, after all, a result of fusion in the deuterium fuel.

Their initially incorrect conclusion had occurred because the neutron diagnostics the ZETA team had used were simply not good enough to tell the different sources of the neutrons apart. While disappointing, this null result, along with the lifting of the lid on the UK's fusion research programme, led to collaborations with various universities as well as improving measurement techniques as a result of working with experts like Mike Forrest.

Perhaps, perhaps, perhaps

While Forrest and his colleagues were working on ZETA, on the other side of the Atlantic, scientists in the United States had been working on 'Project Sherwood', a controlled thermonuclear research programme, since the early 1950s. The US project aimed to create a limitless source of energy by fusing together nuclei of light atoms in a controlled way. This programme of magnetic fusion and energy research was based at the Los Alamos National Laboratory (LANL) in New Mexico. The initially top-secret LANL site had been founded in the 1940s to house the 'Manhattan Project', which succeeded in its aim to create the first atomic bomb.

Project Sherwood got off to a good start. Researchers began experiments in 1951, and just six years later the Los Alamos scientists had succeeded in producing the first ever controlled thermonuclear plasma. This was achieved thanks to a fusion device known as Scylla I, which was a 'theta' pinch device. Like the Z-pinch machine, Scylla I used the magnetic field associated with an electric current flowing through the plasma to 'pinch' the plasma away from the interior walls of the machine. ('Z' and 'theta' just refer to the different directions in which the magnetic field that shapes and confines the plasma acts.)

This type of device has the drawback that the plasma is unstable, and if that instability results in the plasma coming into contact with the machine walls it rapidly loses energy and cools below a useful temperature. Consequently, a range of different configurations were tried out to mitigate this. These included a figure-of-eight-shaped machine at Princeton University designed by Lyman Spitzer, which he named the 'Stellarator'. We will hear more about stellarators in the next chapter.

At Los Alamos, a circular arrangement was tried instead. The device was charmingly named the 'Perhapsatron' and acquired its moniker because perhaps it would work or perhaps it wouldn't. A series of Perhapsatron machines were built to carry out different experiments on the plasmas and magnetic fields used to tame them.

Other devices being experimented with included the Hydromagnetic Plasma Gun created in 1957. This was able to accelerate several litres of hydrogen plasma into a plasma jet moving at 150,000 metres per second (492,000 feet per second or 335,500 mph). Several types of plasma gun were tried out, all as part of research to investigate methods of injecting plasmas into fusion machines.

In stark contrast to fusion research today, which relies on international collaboration, the earliest experiments in fusion energy in the United States – and those carried out by other nations too – were kept close to their respective chests. Although the US declassified much of its fusion energy research in 1958 to facilitate collaboration that could move the research on more rapidly, some aspects of the work were still kept under wraps.

Back in the UK in the mid-1950s, the fusion work was equally under wraps. So, given this background of international secrecy, just how did Mike Forrest and his colleagues end up working in Moscow with Soviet fusion scientists at the height of the Cold War?

To Russia with love?

The initial milestone during the time Forrest was working at the UKAEA was the publication of the *Nature* paper on his game-changing laser measuring technique. This described for the first time how plasma temperature and density could be accurately measured. Forrest and his colleagues soon started to use this technique to also work out the current flowing in the plasma. They could do this because the scattered spectrum gets distorted by the current, so they could work back to figure out where the currents were in the plasma.

Simultaneously, the Soviets had been working on their Tokamak T3 machine and had claimed they were reaching a world-leading temperature of 10 million degrees Celsius (18 million degrees Fahrenheit). This was double the temperature the UK was attaining with ZETA. The Americans hotly disputed the Soviet claims, quite possibly for political reasons since this was a tense period in the Cold War. Or possibly because these results did not meet with anything the US researchers were seeing experimentally or theoretically predicting. Either way, the British fusion experts believed the Soviet results but the Americans did not.

Throughout the Cold War, scientists from either side of the Iron Curtain would meet for the Pugwash Conference on Science and World Affairs, which takes place every five years and seeks to diminish the threat from nuclear weapons. It was at one of these conferences in 1968 that the then director of Culham, Bas Pease, who effectively headed up fusion research in the UK, was asked by his Soviet counterpart Lev Artsimovich, an academician of the Soviet Academy of Sciences, if the UK laser technique could be used to measure the Tokamak T3 temperatures accurately so that no one could be in any doubt over their results. One of the two methods the Soviets had been using to obtain temperature readings was based on extremely complex measurements of plasma energy. The other involved extrapolating the temperature from plasma resistivity measurements. Neither of which were providing a consistent temperature number. Hence the request that the British team behind this new laser measuring technique – Forrest and his colleagues – be sent out to Moscow to carry out some experiments to unequivocally determine the T3 plasma temperatures.

It soon became clear that the Culham management would not have the final say on accepting this invitation. They would have to get Ministry of Science and Cabinet Office approval for the visit. This was because although the teams were studying fusion energy, they contained experts who had been working on each country's respective H-bomb programmes.

Getting spooked

Gaining the security clearance needed on both sides of the Iron Curtain was not aided by the fact Forrest had been assigned to military intelligence during his National Service. By the time of the Soviet invite, he also had close contacts with colleagues working on the UK nuclear weapons programme at Aldermaston. Nervousness was understandable, not least because in the late 1940s Britain's MI5 had discovered a 'mole' working in fusion research in the shape of German physicist Klaus Fuchs.

Fuchs had worked on UK nuclear weapons research before being sent to join the US Manhattan Project to build the first atomic bomb. Returning to the UK in 1946, Fuchs took up a top post at the UKAEA's Harwell site working in energy research. But by 1949, he had been exposed as a Soviet spy. He finally made a partial confession to his activities the following year. Fuchs' crimes, for which he was sentenced to fourteen years in jail, included passing information about the US atomic bomb design to the Soviets.

Against this prevailing background of mutual mistrust, Forrest tells me the Soviet scientists 'had quite a job to convince their people to let me go'. Part of the reason later turned out to be because their tokamak was sited right next to the facilities for the Soviet nuclear submarines. But eventually, despite the obstacles, the ghosts of previous spy scandals were laid to rest and, fortunately for the future of fusion physics, the scientists received Cabinet Office approval for the visit.

Peering under the Iron Curtain

By December 1968, Forrest and his colleague Peter Wilcock were in the Soviet Union on a reconnaissance trip to see the Tokamak T3, which was housed in the Kurchatov Institute on the outskirts of Moscow, to determine if the plasma temperature measuring project was feasible.

They evaluated all of the technical difficulties and soon discovered the T3 machine had tremendous vibration problems and electrical pickup issues. They also sourced engineering drawings to check that they would actually be able to integrate their laser system within the tokamak. After three days studying and evaluating the Soviet device, Forrest and Wilcock concluded they could design their equipment to fit. They had also been required to evaluate whether they would be able to work with the Soviet team. This, Forrest explains, was clearly going to be no problem at all as they 'felt at home straight away' when working with their Russian counterparts.

Forrest and Wilcock duly reported their findings back to the management at Culham, and after a further succession of high-level meetings, the scientific collaboration was finally approved. Now the problem was how best to achieve the aim of the exercise.

Building project

Forrest's team, with the help of colleagues, built the bespoke laser measurement system for the Russian tokamak from scratch in just three months. They knew there was so much electrical noise coming from the Tokamak T3 that alongside their laser equipment they would have to take a so-called screen room (which physicists often refer to as a 'Faraday cage') to house it in. A screen room is essentially a big metal box that prevents external electrical noise from interfering with any equipment placed inside it.

This meant the whole apparatus was so large that only

one airline could fly it – Pakistan Airlines. On 16 March 1969 the airline laid on a special flight for the trip, using a Boeing 707 airliner with big enough cargo doors to get the roughly 5 tonnes of required equipment into the hold. Forrest was sat above the precious cargo in the passenger area, and upon arriving in the Soviet Union he met up with Derek Robinson, a young theoretical physicist from Culham who had also been posted to the Kurchatov Institute.

Forrest and his UK colleagues who flew out later to join him were put up in a guest flat inside the institute, and they were soon hard at work setting up their equipment. It turned out that all of the Soviet scientists spoke English, while the technicians didn't. Despite any language barriers, within a few months the joint team had obtained accurate measurements of the groundbreaking temperatures that the T3 scientists had been claiming all along. The breakthrough came in July 1969 while Forrest had flown back to the UK for a short break. Measurements revealed the T3 plasmas were indeed reaching 10 million degrees Celsius. Fusion research would never be the same again.

A change of direction

A turning point internationally for fusion energy studies had been reached, thanks to this collaborative effort between the two nations. While previously there had been a lot of secrecy even between friendly nations, after the results from this collaboration were published, the veil was lifted on fusion research. It began to become a collective international effort rather than a series of separate, classified projects.

Change in terms of the physics happened even sooner.

Shortly after Forrest and his colleagues returned to the UK, fusion research in the West changed tack. Forrest recalls that as soon as their initial results were in, Culham director Bas Pease had phoned Harold Furth, the fusion director at Princeton University in the United States, who promptly switched their main experimental programme from investigating stellarators (which we will learn more about in Chapter 3) to a tokamak design.

Forrest and his colleagues published the initial results of the measurement experiments on the Tokamak T3 in the 100th anniversary edition of the scientific journal *Nature* on 1 November 1969. A subsequent larger report ('HMSO Report R107') followed after more detailed experiments were completed. This report became a source book for tokamak physics, which was not surprising since these were the first proper measurements that had ever been carried out on a tokamak.

By the early 1970s, most of the fusion work at Culham was also using the tokamak fusion reactor design. ZETA at Harwell was gone, and a new tokamak machine called the Divertor Injection Tokamak Experiment (DITE) had been set up at Culham. Forrest's work on DITE included measuring the plasma current and hence the electric field in the DITE plasmas.

As the scientists pressed on with their experiments, world events were about to take a turn that would result in international teamwork on fusion energy moving to another level.

Clubbing together

In 1973, there was a worldwide oil shortage caused by war in the Middle East. This induced countries without their own oil reserves to increase their research into other forms of energy generation. At that time, Euratom, now a European Union research programme for nuclear research and training but then the atomic energy agency of the European Economic Community (which became part of the EU in 1993) launched a multinational project to build an experimental tokamak fusion reactor. The aim was to demonstrate that controlled fusion reactions could indeed release large amounts of power. They also wanted to develop the technology to get as near as possible to the conditions needed for a viable energy generating reactor.

In 1977, the decision was made for this experimental reactor, the Joint European Torus (JET), to be constructed at Culham in the UK. (The interior of JET's plasma chamber, which is toroidal – hence 'Torus' – is shown in the image below.) The machine achieved its 'first plasma' six years later, on 25 June 1983.

**A view of the plasma chamber inside
the JET tokamak with plasma superimposed.**
UK Atomic Energy Authority/EUROfusion

The next big milestone for JET occurred in 1991, when tritium fuel was added to the plasma. This resulted, briefly, in fusion power of 40kW. As we will explore further in Chapter 3, since that time experiments at JET have been focused on refining tokamak technology to help pave the way for further progress at its successor ITER. The original plan was for JET to run for ten years, but it has instead been in operation for over 30 years and is not scheduled to be decommissioned until 2024. It has lasted so long because it has been repurposed according to the contemporary needs of the fusion community as technologies and fusion research have evolved.

Once it was proved that it was indeed possible to build a torus and magnetically confine a fusion reaction, emphasis then shifted to using JET to investigate how plasma interacts with different wall materials.

While JET is a European project, the rest of the work at the Culham site during the early years of the JET project was part of the UK's national fusion programme. (This changed in 2000 when the UKAEA took over the running of both projects.) The UK work included the development of a spherical tokamak design. The first ever large, spherical tokamak, known as the Small Tight Aspect Ratio Tokamak (START), was built in 1991. This machine was the forerunner of the Mega Amp Spherical Tokamak (MAST), which was constructed in 2000. We will hear about MAST and its successors in the next chapter.

JET proved that controlled fusion reactions that gave out power could be created on Earth. By anyone's standards this was a major achievement. But JET could only manage this feat for a second. In order to use fusion for electricity generation, power needs to be produced continuously. No one has yet demonstrated a sustainable fusion reaction, and this is exactly what is needed for fusion energy generation to become viable. Step forward the ITER project, which aims to create a sustainable reaction for a minute or two and also to demonstrate a viable fuel cycle. Only once this has been achieved can a prototype reactor be created.

International relations

By the 1970s, the international sharing of results between teams working on fusion had become commonplace, but major scale East–West collaboration on nuclear fusion research did not begin until 1985. It was in the November of that year that the President of the Soviet Union Mikhail Gorbachev and the President of the United States Ronald Reagan attended a summit in Geneva, Switzerland, during which they agreed to pool their resources to develop fusion energy. They were pushing for 'the widest practicable development of international cooperation' to reach the goal of fusion energy generation 'for the benefit of all mankind'.

It took 21 years for this to come to pass when the ITER agreement was signed at the Élysée Palace in Paris on 21 November 2006. ITER is the world's biggest international scientific collaboration, and the agreement established 'the ITER International Fusion Energy Organization for the Joint Implementation of the ITER Project'.

As we will see in the next chapter, since the signing of the agreement, a lot of progress has been made on ITER even though the first plasma in the reactor is not scheduled to be achieved for several more years.

Cooling off

So far, the pathway to fusion energy is proving to be a long and complex one, and in 1989 fusion research took a truly bizarre detour. Two chemists, Martin Fleischmann and Stanley Pons, announced in a press conference that they had observed so-called cold fusion in a laboratory test tube. Fleischmann and Pons, at the time based in the chemistry department at the University of Utah, claimed to be able to see hydrogen undergoing nuclear fusion at room temperature rather than at the several millions of degrees found inside a star. Unlike the multi-million-dollar physics facilities experimenting with nuclear fusion, the pair were using nothing more than standard test tubes and readily available laboratory apparatus. The fusion they asserted to be demonstrating was the result of immersing two electrodes made from palladium metal into 'heavy water' and passing an electric current between them. Heavy water is water in which the most commonly occurring form of hydrogen is replaced by the heavier deuterium isotope. Passing the current between the electrodes caused electrolysis to occur. In general, electrolysis results in the electrolyte, in other words the conducting solution that the electrodes are dipped in, to be separated into positive and negative ions. These are respectively attracted to the negatively charged and positively charged electrodes, thereby splitting up the electrolyte. This process is often used for plating objects with metal, as metal ions released from the electrolyte solution are then deposited on the object which takes the place of one of the electrodes.

Pons and Fleischmann claimed that their experiment was splitting off the deuterium ions. These, according to the researchers, were then passing into spaces between the palladium atoms of the electrodes and becoming so densely packed together that they were undergoing fusion into helium. They reported not only large amounts of heat being produced, but also that they had detected neutrons, helium and tritium coming from their apparatus – exactly what you would expect to see if fusion was taking place.

But far from being the discovery of a new, clean energy source that could change the world, as newspaper headlines around the globe starting claiming, this turned out instead to merely be a false alarm.

Within months of the shock announcement, the research had been debunked. Attempts by various other teams of scientists have failed to find any evidence that such a phenomenon exists. As a result, it is surprising then that Google began funding a series of experiments into cold fusion to the tune of $10 million (£7.6 million) in 2015. The Google team planned to develop a rigorous experimental regime for enabling cold fusion, which if detected could then be verified by other scientists. After four years of trying, the series of experiments came to a close and the team published their conclusions in 2019 in the scientific journal *Nature*.

For anyone still clinging to the hope that Pons and Fleischmann might have been on to something after all, it made for disappointing reading. The Google research team had found no evidence for the existence of cold fusion. While this result does not completely close the door on the possibility the effect could occur, it looks extremely unlikely. Not only has there been no experimental proof of the effect, but also there is no current theoretical basis in physics that would explain it either. Thankfully, trying to recreate this fusion-in-a-test-tube has not been a complete waste of effort as it has led to some advances in experimental measurement techniques and materials science. For example, the Google work created some spin-off technologies, including heat measuring devices that can operate under extreme conditions.

But for the time being at least, it looks as though the dream of cold fusion will remain exactly that. Instead, as we will see in the next two chapters, efforts are being focused on a range of fusion reactor designs that all involve incredibly high temperatures.

MAGNETIC ATTRACTION 3

The front runner?

A lot of the experimental fusion reactors in the world today, or currently being built, are based on the tokamak design. This is seen by many as the most likely design for achieving a commercially viable fusion power station. As outlined in the previous chapter, a tokamak is a fusion reactor in which a high-temperature plasma is kept in place via a toroidal magnetic field.

As we have seen, this is the reactor design used at JET and MAST, developed alongside JET at Culham. It is also the technological basis for the international ITER project. At ITER, scientists plan to maintain fusion for long periods of time and test out the integrated technologies, materials and physics needed to move to commercial production of fusion-based energy. As we will see shortly, there are also private companies working on tokamak-based designs, while stellarators are also holding promise for energy generation.

To picture what a tokamak looks like, imagine the

following fantasy scenario. You go to the bakers and buy a doughnut, and then for reasons best known to yourself decide to have it gold-plated. You now have a doughnut with a metal casing. Now, imagine that you drill a hole in the bottom of the gold-plating and heat the whole thing up so high that the doughnut melts and pours out of the hole.

This is essentially the basic structure and operation of a tokamak fusion reactor. The doughnut is the plasma, the inside edge of the gold-plating is the plasma-facing wall and the drilled hole at the bottom is the equivalent of a component known as the divertor, through which spent fuel flows out and new fuel is injected in.

The world's largest collaboration

Of course, a real reactor is somewhat bigger than a doughnut. ITER's 23,000-tonne tokamak, which is currently under construction at a 180-hectare site in Cadarache, will be the largest ever built. It will have an internal volume sufficient to house 830 cubic metres (219,000 US liquid gallons) of plasma, dwarfing the 100-cubic-metre (26,000 US liquid gallons) plasma volume of the largest tokamaks to date – those of JET in the UK and JT-60 in Japan. It is the large scale of ITER's plasma volume that will allow it to create a so-called burning plasma. This is a plasma which is past its ignition point and so heated mainly via the alpha particles created in the fusion reaction, rather than solely via external heating.

At ITER, the aim is to create the first fusion device to produce net energy. The plan is to create ten times more power than has to be put in to heat the plasma. So, 50 megawatts (a megawatt is 1 million watts) of power will go in,

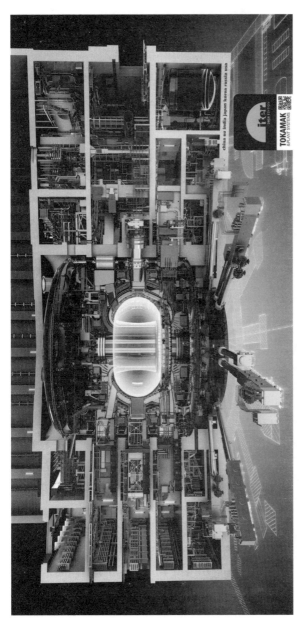

This cut-away illustration shows the geometry of ITER's tokamak and the shape the plasma will be once the machine is operational.

ITER Organization

but 500 megawatts of thermal fusion power will be produced. That said, this power output will only last for 400 to 500 seconds, with an ultimate goal of maintaining about 350 megawatts of output for almost an hour. Although achieving this would be an important milestone, it helps underline just how challenging the move to commercial energy generation will be.

Despite the vast scale of the project, ITER cannot achieve everything required to move to a fusion-powered future on its own. It is being helped by a 9.4-metre (30.8-feet) diameter 9.6-metre (31.5-feet) high, 1,000-ton pilot device based in Daejeon in South Korea known as KSTAR. KSTAR's main mission is to achieve high-performance and steady, continuous operation of its tokamak fusion reactor. This is in no small part assisted via its essential and innovative superconducting magnet system, which we will learn more about in Chapter 5.

KSTAR first got the go-ahead in December 1995, and by the August of 1998 the design concept and basic research and development phase had been completed. It was then a case of designing the facility from an engineering perspective, which took until May of 2002. By August 2007, construction was complete and less than a year later, in June 2008, the first plasma was achieved. The facility began full-scale experiments the subsequent year, and in 2010 began clocking in a succession of world-firsts.

The initial achievement was reproducing high-performance plasma operation conditions in a superconducting tokamak, then successfully extending the duration of those conditions. By 2012, the KSTAR team were able to hold the high-performance plasma operation conditions for seventeen seconds, and six years later they had extended this to 90 seconds.

In addition, KSTAR managed to reach a plasma ion temperature of 100 million degrees Celsius (180 million degrees Fahrenheit). This latter result, along with other achievements in operating the technology, forms part of KSTAR's third stage of operation, with the fourth and final stage due to start in 2023 and run for roughly two years. This last chapter aims to hold the plasma in KSTAR at 100 million degrees Celsius for 300 seconds by the year 2025. (While in large scale fusion reactor designs the ideal operating state for the plasma is for the ions and electrons to be in thermal equilibrium, in other words at the same temperature, smaller systems have ion temperatures hotter than electron temperatures and others are vice versa. This is because in the former case their heating method preferentially increases the temperature of the ions, while in the latter case the electrons are heated more strongly. In KSTAR, the ion temperature is above the value of the electron temperature.)

In November 2020, KSTAR set a new world record in maintaining plasma hot enough to sustain fusion power. It held the ion temperature of the plasma at 100 million degrees Celsius for twenty seconds, doubling the previous duration record. This was no mean feat, given that this temperature is several times hotter than the centre of the Sun, and was twelve seconds longer than KSTAR had managed just the year before. In fact, the 2019 progress was relatively swift too, given that in 2018 this ultra-high temperature could only be maintained for 1.5 seconds.

Meanwhile, at the Southwestern Institute of Physics in China a medium-sized tokamak known as HL-2M has been running experiments designed to support the physics and techniques that need to be employed in both ITER and the forthcoming Chinese Fusion Engineering Test Reactor.

The design of HL-2M allows, for example, different types of divertor configurations to be tried out. The external magnets are also removable, which enables improvements to be made to various machine components as more is learned. In December 2020, HL-2M made headlines when it was powered up and reached temperatures above 150 million degrees Celsius (270 million degrees Fahrenheit).

Jetting off to a good start

None of this work would have been possible without the forerunners we learned about in the last chapter, including JET, based at the Culham Centre for Fusion Energy in the UK. Building work started on this tokamak in 1978, and it was operating five years later with the first plasmas being created.

In 1991, the JET scientists began running experiments with the deuterium–tritium fuel mix that ITER will use. Within months, they had succeeded in achieving the first ever controlled release of fusion power with this fuel combination. This was an important milestone not only for ITER, but also for future commercial fusion plants based on tokamak technology.

JET is currently the world's largest and most advanced tokamak. It is helping pave the way for ITER by not only being used to study plasma reactions and the deuterium–tritium fuel mix, but also by testing out various systems and materials that will be used in ITER. For example, as we will learn in Chapter 6, between 2009 and 2010, the original inside surface of JET's reactor wall was replaced by a new plasma-facing wall made from beryllium and tungsten to see how well these materials could work for ITER.

At the time of writing in 2022, new experiments with an ITER-like deuterium–tritium fuel mixture have recently been completed at JET. They provided a 'dress rehearsal' for ITER by sustaining high fusion performance for longer periods. It is fair to say that the results could not have been much better. The JET machine released a record-breaking amount of sustained fusion energy – 59 megajoules. In terms of energy alone, this was more than double the 22 megajoule previous record set by JET, which was achieved back in 1997. But on top of that, this 2022 result came as very welcome news for ITER as it was the first ever demonstration of a sustained plasma using the very same wall materials and fuel mixture, thereby confirming that these choices were leading ITER along the right track.

Culham is also home to the Mega Amp Spherical Tokamak Upgrade experimental fusion device, which, like JET, is testing out physics and technology for ITER. The original MAST machine ran from 2000 to 2013, and the Upgrade machine is based on this first machine with some improved features. These include a stronger magnetic field, greater heating ability and a new type of divertor – which is the plasma exhaust system for a tokamak.

As Chapter 5 will reveal in more detail, a divertor has a lot to cope with since the extremely hot spent fuel from the reaction passes out through it. In fact, the hammering a divertor takes from the amount of heat and power it has to withstand would mean that commercial fusion power plants would have to replace divertor components every few years. This process would be so costly and time consuming that it would raise a question mark over the viability of using fusion power as an electricity source. Therefore, the MAST Upgrade machine will be the first tokamak to be fitted with a 'Super-X divertor', which is specially designed to reduce the heat and power loads coming from the particles exiting the plasma. This should prolong the life of the divertor components. But if Super-X does not work as well as expected, other divertor designs can be tried out with the MAST Upgrade so that the optimum design for the power plants of the future can be found.

It is also hoped that the spherical shape of the MAST Upgrade, with its more compact magnetic field compared with JET and other conventional doughnut-shaped tokamaks, proves to be a design that moves us nearer to the commercialisation of fusion. In this case, this would be via smaller fusion power plants that would be cheaper to build than those based on the ITER design.

The next phase for UK research will be the construction of the Spherical Tokamak for Energy Production (STEP). Once operational – scheduled for the 2040s – this tokamak will act as a prototype for a working power station and will be connected to the UK's National Grid. It will test out components, materials and robotic solutions, as well as computer modelling. The plan is that the conceptual design for the reactor will be finished in 2024, and at the time of writing, five potential sites for housing this prototype are being assessed.

Not to be overlooked are the experiments that were carried out in the United States at the Tokamak Fusion Test Reactor (TFTR). This tokamak was based at the Princeton Plasma Physics Laboratory and ran from 1982 to 1997. While JET had briefly introduced tritium as fuel in 1991, TFTR became the first magnetic fusion device to carry out comprehensive experiments using the mix of fuel that will be needed for commercial tokamak power plants in the nominal range

A conceptual illustration of the Spherical Tokamak for Energy Production plant. The tokamak chamber shown on the front right is partially underground.
UK Atomic Energy Authority/EUROfusion

of a 50/50 deuterium/tritium mixture in 1994. (In fact, since tritium is regulated as a nuclear material, JET and TFTR are the only two machines to have been licensed for D–T fuel so far.) However, TFTR was not designed as a high-performance tokamak. Therefore, the subsequent experiments at JET in the late 1990s eclipsed TFTR's results and, as we have just seen, are leading the way again now.

Private investigators

Private companies are also developing reactors based on magnetic confinement technology. At the time of writing, investment into these companies, and into private companies working on other reactor designs (some of which will be detailed in Chapter 4), is growing at a fast rate. But it is worth bearing in mind that a sector working on an evolving technology is, by its very nature, unpredictable. Only time will tell if one or more of the technological advances planned by, or already achieved by, these companies will play a significant role in providing the world with fusion energy.

With so many companies in this growing sector it would be overwhelming to look at all of them. But in this chapter and the next we will get a flavour of what fusion reactor concepts are being considered and experimented with and of how some of these companies are hoping to progress.

As we will explore later, it is looking as though no single player will solve all the problems on the route to achieving viable fusion energy generation. Instead, that success is most likely to be achieved by combining technologies from publicly funded projects and private enterprise. What is clear is that the private fusion sector is now attracting serious levels of interest from investors, and so has its own commercial imperatives and drivers.

Magnetic attraction

One private fusion company using a tokamak, but in this case a spherical one in the shape of a cored apple, is Tokamak Energy based in Oxford in the UK. Having a reactor this shape makes it possible to achieve a much higher plasma pressure for the same value of magnetic field compared with a doughnut-shaped tokamak. Because the magnets used to confine plasmas are so expensive, this arrangement keeps costs down while still offering high levels of power output.

Another point of difference compared with tokamaks of the ITER type are the high-temperature superconducting magnets that Tokamak Energy is using. These magnets are made from rare-earth barium copper oxide (REBCO) which can operate between –250 and –200 degrees Celsius (–418 and –328 degrees Fahrenheit), rather than –269 degrees Celsius (–452 degrees Fahrenheit) for more conventional superconducting magnets. They are very effective at containing the plasma and are much more compact than other types of superconducting magnet, helping to keep the overall size of the reactor relatively small.

In 2022, the company is aiming to demonstrate a reduced-scale version of their toroidal (doughnut-shaped) and poloidal (running in a circle around the circumference of the reactor) magnetic coils made from REBCO. This magnet design is destined for Tokamak Energy's planned fusion pilot plant. Tokamak Energy's existing ST40 reactor uses copper magnets and stands approximately 4 metres (13 feet) tall and 3.5 metres (11.5 feet) across and contains a plasma around 1.3 metres (4.3 feet) wide. Although Tokamak Energy's ultimate objective is a commercial fusion power plant, one of their next targets is to achieve a 100-million-degree plasma temperature in ST40.

Variations on a theme

As we have just seen, magnetic confinement for fusion does not have to involve a conventional tokamak design, and variations on this technology have been chosen by several private fusion companies. These include Seattle-based CTFusion, which is looking to reduce the cost of fusion reactors by simplifying their design. CTFusion's concept, developed in conjunction with the University of Washington, is to magnetically confine fusion plasmas in a design known as a spheromak.

This is a cousin to a tokamak and uses magnetic fields from electric currents flowing inside the plasma to confine it, rather than using superconducting magnetic coils outside of the reactor chamber to create confinement. The spheromak still has a single set of external coils to keep the plasma stationary for continuous operation. But one set of coils rather than the three needed for a tokamak makes the spheromak design not only much simpler in comparison, but also substantially cheaper to build. This raises the possibility of siting power plants near to urban centres, thereby minimising the grid infrastructure required to deliver the fusion-generated electricity to homes, businesses and transportation networks.

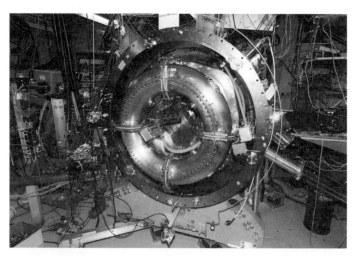

The 0.76-metre (2.49-feet) diameter helicity injection manifold for CTFusion's sustained spheromak (Dynomak™) fusion energy prototype. This component is a critical part of their unique plasma current drive technology.

CTFusion, Inc.

In the CTFusion system, the electric currents that generate the magnetic fields to confine the plasma also heat the
plasma to the temperatures required for fusion. This dual use
for the current simplifies the design of the reactor compared
with a tokamak, as well as helping to keep it smaller. In 2014,
the company patented their unique Dynomak™ technology
approach to the spheromak confinement concept, which
includes a method of sustaining the plasma current in a very
efficient way. This concept is based on the same physics that
produces prominences from the surface of the Sun, CEO of
CTFusion Derek Sutherland explained to me.

Underneath the Sun's surface, electric currents flow that
generate twisted magnetic field lines that extend outwards
into space. Plasma from the Sun then flows along these magnetic field lines, forming the red, glowing loops known as
prominences which can be hundreds of thousands of miles
in length.

'Our insight was [to ask] what if we create these types
of prominences artificially in the lab and use them to drive
currents in a machine?' said Sutherland. CTFusion demonstrated that this worked in their first Dynomak prototype at
the University of Washington in 2012. Having shown the
concept to be successful on a small scale, the company began
working towards commercialisation of their technology.

'Our modelling and simulation suggests an increase in
size will improve plasma performance considerably, but of
course we need to confirm it experimentally,' said Sutherland.
So, the next step on the journey towards net energy gain is
for CTFusion to build a larger device and move to higher
performance plasmas, meaning higher temperatures, densi-
ties and energy confinement times. They will focus on the
most difficult challenge first, which is to make sure that

the physics will scale, and therefore that net energy can be created via the design. Then they will need to prove that they have good enough confinement for the plasma, and a reactor wall design with a suitable lifespan for a commercial power plant.

If their technology proves viable, CTFusion intends to take a modular approach to the engineering. Making the machines in sections would enable multiple fusion reactors to be manufactured simultaneously, and for several reactor units to be linked in parallel, or larger versions of the reactor made, depending on the power requirements in a given location.

Meanwhile, UK-based Fusion Reactors Ltd, founded in 2019 by plasma physicist Dr Christos Stavrou, is focused mainly on working out the optimal magnetically confined fusion reactor configuration by evaluating different reactor designs. By studying each approach, Stavrou, who has previously worked at JET as well as in the private fusion sector, is hoping to quickly discover the optimum design for commercial energy generation, which could be a combination of methodologies.

Not only are the reactor designs produced by Fusion Reactors different from those being used elsewhere, the company has also developed its own unique method for validating these designs rapidly, and at low cost. Stavrou believes 'that slight incremental improvements to existing designs are unlikely to lead to the step change in performance that is needed quickly enough'. With his company's process, 'if a design does not produce the correct performance, we are free to drop it and test another', he explains, adding that this approach makes sense because 'nobody knows yet which design[s] will be able to deliver commercial fusion'.

That said, Stavrou intends on sticking with magnetic confinement fusion because of the large body of research on such systems. At the time of writing, the company is set to start seeking investment and is aiming to 'demonstrate a clear path to commercial fusion within three years from start of operations', he says.

Shrink to fit

Just by looking at the designs proposed by this selection of private fusion companies, it has already become clear that not everyone working in fusion research envisions future fusion reactors in power plants being gargantuan in scale. In fact, some private companies are looking to go very small scale with their fusion reactors. Princeton Fusion Systems, based in New Jersey, for example, has come up with a concept they are calling a 'microreactor' that would be completely portable via a large truck. Their design would consist of a sealed unit with enough deuterium–helium-3 fuel to provide power for 30 years.

Another compact design is being worked on by US aerospace and defence giants Lockheed Martin. Their system would also be able to fit on a truck and could power the needs of a small city of 100,000 residents. The small size of their reactor makes it much quicker to build than larger fusion reactors, which will potentially speed up testing cycles and help the company get to the prototype stage rapidly. This is because they can try out a range of design choices readily, while keeping costs comparatively low compared with larger-scale projects.

The turn of the screw

Another design of fusion reactor based on magnetic confinement which could play a major role in fusion energy development is the stellarator. As covered in Chapter 2, stellarators were invented by the American physicist Lyman Spitzer at Princeton University in 1951. They are similar to tokamaks in the sense that the plasma is controlled by a magnetic field. But in the case of a stellarator the twisted magnetic field is generated solely by external magnets.

To visualise the shape of a stellarator plasma, imagine someone has taken a doughnut-shaped tokamak plasma and twisted it round as though they were wringing out a wet dishcloth. The plasma-facing walls of a stellarator echo the geometry of the plasma they contain. This twisted shape makes for a challenging construction project. However, the positive trade-off is plasma that is more stable and simpler to control than that in a tokamak.

There are several stellarators currently operational, including the Large Helical Device at Japan's National Institute of Fusion Research, the National Fusion Laboratory of Spain's TJ-II, the Helically Symmetric Experiment stellarator in the University of Wisconsin-Madison in the United States and the Wendelstein 7-X at the Max Planck Institute for Plasma Physics in Greifswald in Germany. The latter, which was built between 2005 and 2014 and produced its first plasma in December 2015, is the largest working stellarator in the world.

One of the main technical challenges for stellarators is that they leak both heat and particles from the plasma they confine, resulting in energy losses. This problem is so bad that it has hampered the development of stellarators. But research

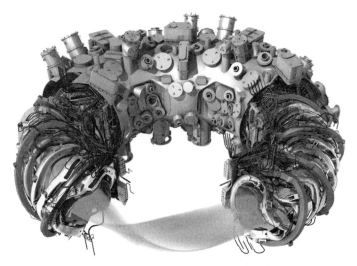

Diagram of the world's largest working stellarator, the Wendelstein 7-X stellarator at the Max Planck Institute for Plasma Physics in Germany.

Max Planck Institute for Plasma Physics

published in the scientific journal *Nature* in 2021, by a collaboration of scientists from the Max Planck Institute, the National Fusion Laboratory of Spain in Madrid, the Laboratory for Plasma Physics in Brussels and the US Department of Energy's Princeton Plasma Physics Laboratory, showed that careful alteration of the magnetic field confining the plasma could reduce the amount of leakage, thereby offering the tantalising prospect that the stellarator design might have what it takes to become the basis for a power plant.

The magnetic field for the Wendelstein 7-X stellarator that was used for this demonstration was created thanks to 50 superconducting niobium–titanium magnet coils. Each coil consists of niobium–titanium strands embedded in copper that are twisted to form a cable that is enclosed in an aluminium sheath. To cool the coils to temperatures low enough for the niobium–titanium to become superconducting, liquid helium is pumped around a kilometre-long pipe system that delivers it in between the wires and the sheath. During the design phase, the optimum shape and position of these coils had been calculated using a supercomputer to provide the best possible confinement of the plasma in the reactor chamber of the stellarator.

Also in 2021, scientists working on the Wendelstein 7-X stellarator made further progress in their quest to discover if this type of fusion reactor is a viable design for a future power station. They installed a new water-cooled divertor that is so efficient at flushing heat out of the stellarator that they can now run the machine with plasmas that last for 30 minutes. This opens up the opportunity for experiments to be carried out in conditions that more closely mimic a potential commercial fusion power station and moves closer to the continuous operation that would make a stellarator attractive for energy generation.

Smaller wonders

There are also some much smaller-scale stellarators being developed or experimented with. Private company Renaissance Fusion based in Grenoble, for example, is aiming to build the first stellarator with walls made from flowing liquid metal. They are working in collaboration with several partners, including the National Institute for Research in Digital Science and Technology in France.

Type One Energy Group are also developing a net power

stellarator design for their STARBLAZER I machine. This private company, based in Wisconsin, is seeking to optimise the magnetic confinement of the plasma using high-temperature superconducting magnets to help reduce the overall size and cost of their reactors. Once they have determined the most effective magnetic field configuration, Type One Energy, which are working in conjunction with various partners including Massachusetts Institute of Technology, Commonwealth Fusion Systems and the Oak Ridge National Laboratory, intend to construct STARBLAZER II, a commercial demonstration reactor.

Focusing in

LPPFusion is taking another approach. As with a stellarator, they are magnetically confining a plasma in their design, but in their case this is not via the use of external magnets. Instead, they are using the magnetic field from an electric current to confine their plasmas via a fusion method known as Dense Plasma Focus. This reactor design is based around two cylinder-shaped metal electrodes, one nested inside the other within a vacuum chamber that is filled with a low-pressure gas and would in a commercial reactor also contain the fusion fuel. These electrodes are relatively small, with the outermost generally measuring just under 18 centimetres (7 inches) in diameter and slightly over 10 centimetres (4 inches) in length. In the case of LPPFusion, the fuel they intend using in their 'Focus Fusion' technology is proton–boron (pB11).

Within the Dense Plasma Focus device, a short pulse of electricity is used to create plasma from the gas in the vacuum chamber. This pulse is sent across the electrodes, which for a few millionths of a second causes a large electric current to flow from the outer electrode to the inner electrode by passing through the gas. It is the gas in the path of this current which is turned into plasma.

The magnetic field from the electric current confines the plasma, pinching it down into tiny plasma filaments and twisting them around until a dense ball of plasma called a 'plasmoid' is formed that lasts for just 10 nanoseconds (10 billionths of a second). The friction from the electrons moving through these plasma filaments then heats the plasmoid up to an extremely high temperature – billions of degrees Celsius or more – in a similar way to how the flow of electrons in a lightning strike heats up the air it passes through.

Diagram of plasma filaments in the LPPFusion system.

Courtesy of Lawrenceville Plasma Physics Laboratories, LPPFusion.com

Fusion occurs in the plasmoid, and instabilities in the plasma lead to the creation of an electron beam coming out from one direction and an ion beam emerging in the opposite direction. Since a lot of the energy from the fusion reaction ends up in this ion beam, the LPPFusion design involves generating electricity directly from the beam. The LPPFusion team intend to achieve this by using a reversed version of the technology that uses electricity to accelerate charged particles in a particle accelerator. In their system, they will decelerate the charged particles in the ion beam to create electrical power.

The company, which was founded by Eric J. Lerner in 2003, is already producing fusion reactions in their reactor and will move on to testing with their pB11 fusion fuel once they have completed the test schedule for their main reactor systems. Their aim is to have a pilot reactor demonstrating the commercial feasibility of this method of fusion by 2025.

Given the amount of time, expertise and financial investment going into ITER and other fusion projects using magnetic confinement, it might seem strange that alternative fusion concepts are still being investigated. But fusion is a complicated business. And since no reactor design has yet proved itself fully, in the meantime, various institutions and companies are looking at other approaches to fusion energy, as we will see in the next chapter.

COMPETING TECHNOLOGIES

4

Beam us up

Since it is too early to tell if magnetic confinement will definitely prove to be the solution for the world's first fusion power station, research into alternative fusion technologies is continuing apace. The principal alternative to the magnetic confinement systems we have just heard about is inertial confinement, which as we will see shortly is already used in nuclear weapons research.

In this type of fusion, either a laser or a particle beam is fired towards a target of hydrogen fuel contained in a small pellet. The pressure generated by rapidly ablating (in other words, removing material from) the surface of the capsule crushes the fuel so much that it briefly creates a plasma in which hydrogen nuclei are pushed into each other and fuse.

Inertial confinement is the approach being taken at the 'Z' machine at the Sandia National Laboratories in Albuquerque, New Mexico, where work on inertial confinement fusion energy research takes place alongside work on nuclear weapons.

In Sandia's magnetised liner inertial fusion (MagLIF) fusion energy experiments, the Z machine focuses incredibly large electric currents – around 26 million amps – for a few nanoseconds onto a 'liner', which is a metal cylinder filled with fusion fuel. The extremely large magnetic field produced by these huge electric currents crushes the liner and the compression turns the fuel into plasma in which fusion takes place.

Operating this device is quite dramatic. When the Z machine fires, it creates over 1,000 times more electricity than in a lightning bolt and so much vibration is caused that glasses of water would be shaken in buildings several

Dr Matthew Gomez, the lead physicist for the MagLIF fusion effort at Z standing in the Z arena. Behind him are sections of the huge laser that is part of Z's fusion effort.

Courtesy of Sandia National Laboratories

hundred metres away. For generating electricity, the aim would be to extract as much energy as possible from the very small but high-density plasma created via each firing of the machine before that plasma expands outwards and cools.

The promise for the MagLIF 'magnetic direct drive' concept lies in the fact that magnetising and pre-heating the target enables the fusion fuel to produce a higher fusion energy yield than it otherwise could.

Following the success with this approach at the Z machine, the National Ignition Facility (NIF), based at the Lawrence Livermore National Laboratory in California, began working on using magnetisation to improve the energy yields from their fuel capsules. The NIF, which began operating in 2009, focuses pulses that have been amplified 1 million billion times from 192 laser beams onto a pencil-eraser-sized target holding a fuel pellet within a 10-metre (33-feet) diameter target chamber. These fuel pellets contain deuterium and tritium. The lasers are so big they need to be housed in a ten-storey building the width of three American football pitches, and when their beams converge on the hohlraum (the gold canister that holds the target) they generate X-rays. These X-rays cause the target to implode, generating massive pressures and temperatures that fuse the hydrogen atoms.

In fact, the fuel pellets are put under pressures more than 100 billion times greater than the atmospheric pressure we feel every day. These are similar pressures to those found inside the Sun. So, the close-to 2 million joules of ultraviolet laser light delivered causes the pellets to produce a 100 million-degrees Celsius (180 million degrees Fahrenheit) plasma. The fusion reactions that then occur within this plasma release several types of particles, including alpha particles.

When the alpha particles react with the plasma surrounding them, this heats up the plasma more, creating additional fusion and hence more alpha particles, and so the cycle goes on. If this self-sustaining reaction produces more energy than is supplied by the laser, it has reached 'ignition'. It is critical to reach this point, because this is the stepping-off point to higher energy gain and the tantalising prospect of net energy generation.

No experiment has approached this level of self-sustained fusion. But on 8 August 2021, an experiment at the NIF came closer than any have before and achieved a yield of 1.35 MJ. This was eight times their previous record for energy output, and since they put 1.9 MJ of energy in to begin this experiment, this yield was equivalent to 70 per cent of the laser energy fired on the target.

While this was a major breakthrough for the NIF team and their collaborators, overall, they are still putting more energy in to create fusion conditions than they are getting out from the fusion reaction. Therefore, the next milestone will be to achieve a break-even point, where the amount of energy coming out exceeds that going in. The data from those experiments should reveal whether inertial fusion energy is a viable technology for future power plants.

As with the Z machine, the quest for fusion energy is not the sole thrust of the research at the NIF. Instead, one of its main functions is providing the classified experiments needed to keep the United States' stockpile of nuclear weapons safe. But those same NIF methods that help maintain the safety and reliability of the US nuclear deterrent can equally be used to recreate the conditions found in the interiors of stars, planets and other objects in the universe. So, alongside research into fusion for future energy generation, the facility is also carrying out experiments designed to increase our understanding of the fusion process in stars and black holes.

Guiding lights

None of the experiments at the NIF would have been possible without a lot of work developing bigger and better lasers. These have now evolved through several generations. The first of the Lawrence Livermore National Laboratory's large laser systems was known as Janus. This was built in 1974 and consisted of a single beam. While this might sound very small compared with the NIF's current 192-beam laser system, in its time, Janus was groundbreaking. It enabled the first ever inertial confinement fusion reaction to occur by imploding fuel pellets of deuterium–tritium. But before the end of its debut year, Janus had been expanded to a two-beam system that helped scientists understand more about the fusion reactions it was initiating.

In tandem with Janus, another single-beam laser known as Cyclops was coming online. This was to provide a test bed for the twenty-beam Shiva laser completed in 1977. Shiva also had design input based on the NIF's experiences with their 1976 two-beam system known as Argus. The development of Argus was an important step for laser-based inertial confinement fusion. This was because it helped prove technology that enabled the beam to be relayed between amplifiers – thereby increasing its power and energy to the required level, while at the same time not amplifying unwanted fluctuations that could damage the laser components.

It is not only large institutions that have been using

lasers to induce fusion. Laser fusion of hydrogen and boron-11 is the approach being taken by Australian company HB11 Energy, which is working to achieve proof-of-concept for their method. This will involve directing a high-energy laser pulse just a trillionth of a second long towards a pellet of hydrogen and boron-11 fuel. This will result in a plasma, which will be magnetically confined within a small reaction unit to create the optimum chance for the hydrogen fuel to fuse with the boron. This fusion reaction produces helium nuclei, which as we saw in Chapter 1 are alpha particles.

One major advantage of this 'HB11 fusion' is that it does not require the enormous temperatures needed for triggering fusion in tokamaks or stellarators. The reaction does not create neutrons either, so the components for HB11 reactors would not become radioactive, like they do in many other reactor designs, which is a problem we will explore further in Chapter 5. Also, the fuel is easier to handle and acquire since, unlike tritium, neither hydrogen nor boron are radioactive, plus boron is abundant in nature and can readily be mined at scale.

Once fusion is triggered, HB11 Energy plans to create electricity directly from the reaction by giving the spherical wall of the reactor vessel an electric charge. This charge will first slow down the positively charged alpha particles, then as they make contact with the reactor wall capture their electric charge to make electricity from it.

Without the need for steam turbines or generators, reactors based on this concept could be smaller than tokamaks. They are likely to measure 'somewhere between a football field and a shipping container. The size will depend on how lucky we are with the science, which will dictate how big the lasers we need are', Dr Warren McKenzie, managing director of HB11 Energy and adjunct lecturer at the University of New South Wales, explained to me. In contrast to deuterium–tritium fusion, the company believes its concept will allow cheaper and faster development of fusion for electricity generation as well as for space propulsion.

Raising the pulse rate

Continued developments in laser technology as well as in materials science are benefiting private company Marvel Fusion, founded in 2019, and based in Munich. They are firing an extremely short, high-intensity laser pulse at the fuel to trigger fusion. However, their system differs from other inertial fusion set-ups thanks partly to the use of shorter laser pulses. In addition, they have created a specific structure at the nanoscale for their fuel pellets. This enables the fuel to absorb almost all of the laser energy, and for quantum-mechanical effects and other effects to occur that increase the probability of fusion and ignition. Overall, Marvel Fusion's approach enhances the amount of fusion that can be produced from a given amount of their proton–boron-11 (pB11) fuel.

The fusion reaction between protons and boron-11 nuclei releases three helium nuclei (alpha particles) and very few neutrons. Fewer neutrons means less radioactivity created inside the reactor, reducing the need for long-term handling of radioactive components and keeping disposal of radioactive waste to a minimum. (Although fusion does not itself produce radioactivity, the neutrons created in the process can make some of the materials a fusion reactor is constructed from radioactive. We will look at this in Chapter 5, and at the challenges involved in handling and storing radioactive waste in Chapter 6.)

Obtaining charged alpha particles as the result of the fusion reaction has another advantage in that their electric charge could be directly converted into electricity via electromagnetic induction and/or electrostatic convertors. Any process that enables direct conversion to electricity would likely require less space and facilities than steam-driven turbines in a future fusion power station.

Unlike in tokamaks or stellarators, in laser-driven inertial fusion energy, the reactions only occur for a split second as each laser shot hits the fuel. In order to be able to produce energy, laser pulses would need to be repeated extremely quickly with a regular frequency, such as, for example, ten times per second. Many of Marvel Fusion's planned research efforts are looking to develop state-of-the-art fast-pulse laser systems to achieve that aim. If this approach to laser-driven inertial fusion electricity generation becomes commercially successful, fusion plants based on this technology would have the advantage of being able to respond quickly to peaks and troughs in demand on the electricity grid, as when demand is low they could simply induce fewer fusion reactions.

Marvel Fusion's ultimate goal is to build and operate commercial fusion power plants in the 2030s. They are planning to upgrade an existing laser system to carry out a series of experiments through 2022–5 to validate their innovative technological approach and demonstrate the route needed to achieve net energy gain via this method. Alongside these experiments, Marvel Fusion is already working with partners from industry and science to develop the technologies and system components that will be needed for a demonstration plant and for future commercial fusion power plants.

Ready, aim, fire!

Another approach to pulsed fusion is using a plasma gun. This might sound like something out of a 1970s science fiction film, but it is a reality for the HyperJet Fusion Corporation based in Virginia. HyperJet and sister company NearStar Fusion, the latter of which was co-founded by Christopher Jay Faranetta with Douglas Witherspoon and Randy Roy in 2021, both base their approaches on the imploding of fusion fuel targets to create fusion. NearStar Fusion uses metallic 'liners' to contain its fusion fuel. These liners are metal cylinders filled with gaseous fusion fuel. Meanwhile, HyperJet Fusion Corporation, in a research partnership with the Los Alamos National Laboratory, is studying the feasibility of using repetitive imploding liners of plasma to ignite deuterium–tritium plasma fuel with the Plasma Liner Experiment located at Los Alamos.

In a NearStar Fusion power station, fusion neutrons would be produced via repetitive pulses of plasma fireballs. Pulsed operation makes the plant much simpler and could dramatically reduce the amount of power required to either start up or maintain operations compared with fusion technologies that confine plasma in a steady state. Operating in a pulsed manner without the need for magnetic confinement also facilitates the optimisation of capturing the energy released by each fusion pulse.

This pulsed regime is created by what NearStar Fusion term their 'hypervelocity launcher' – aka a plasma armature railgun. Each pulse of fusion energy in the machine is triggered by the firing of a deuterium–tritium fuel capsule out of the hypervelocity launcher. The fuel capsule speeds into an electromagnetic coil, and the magnetic field this coil produces compresses the fuel capsule so much that it ignites and forms a plasma. The heat from the intense burning of the plasma fireball is absorbed by molten salt that flows down the walls of the reactor chamber. In a commercial power plant based on this design, the molten salt would be used to boil water and the steam that this creates would drive a turbine to generate electricity. This approach, explains Faranetta, builds on the science of the Sandia Laboratories' Z machine and has the potential to evolve to burn advanced fusion fuels that create few or no neutrons.

The HyperJet Fusion Corporation power plant concept makes use of up to 600 plasma guns, some firing deuterium–tritium fuel plasma and others firing a plasma liner into a spherical reactor vessel. In this concept, the fuel is fired in first, while the liners follow, forming an imploding shell around the fuel which then implodes. The compression from the liner should be enough to cause the plasma fuel to achieve fusion ignition. Liquid flowing through the reactor chamber walls would then collect the heat from the resulting pulsed fusion reactions.

Faranetta says that they are looking to develop 'each concept as far as it will go towards a practical reactor'. In the future, commercial power plants based on the NearStar Fusion design could be made in a modular way so they could readily be altered or expanded. This modular design would also facilitate easy mass production of the plant components. In terms of materials, NearStar Fusion intends using existing technologies and materials to avoid complex and expensive materials research.

A hybrid approach

There are also private companies integrating pulsed fusion with other fusion technologies. For example, General Fusion, founded by Canadian physicist Dr Michel Laberge almost twenty years ago, is developing a hybrid approach it calls Magnetized Target Fusion (MTF) technology.

This design involves first filling a rotating tank with liquid metal. As the tank spins round, centrifugal force forms a cavity in the liquid metal, into which a magnetised hydrogen plasma is injected. The plasma is compressed by using high-pressure gas pistons to push on the liquid metal walls that surround it. As the liquid metal walls collapse, the plasma briefly heats to more than 100 million degrees Celsius (180 million degrees Fahrenheit), enabling its hydrogen atoms to fuse into helium and release energy. The liquid metal wall captures the heat from the fusion reaction, and in a commercial plant this heat would be extracted for producing electricity.

This is a pulsed approach to fusion because it works in cycles, with the end of a cycle occurring once the liquid metal wall has completely collapsed. Because there is no need for exotic lasers, complex plasma control systems or expensive superconducting magnets, plus the liquid metal liner shields the reactor structure from neutrons released by the fusion reaction, General Fusion believes their technology will be cost-effective and easy to implement. Not least because the required components can be manufactured via existing industrial processes.

Technicians work on a key component of General Fusion's Magnetized Target Fusion technology – the compression system.
General Fusion

Construction of General Fusion's MTF demonstration plant at the UKAEA Culham Centre for Fusion Energy near Oxford is anticipated to begin in 2023, with operations commencing approximately five years later. This location, according to General Fusion, will enable them to benefit from the expertise of the UKAEA teams who have built and run fusion machines, including JET, as well as providing access to existing supply chains necessary for commercialising fusion.

The building project itself looks set to be innovative, with the company wanting to showcase how sustainable industrial design – in the form of materials such as low-carbon concrete and efficient building techniques – can transform the energy sector. To help achieve this, General Fusion has commissioned architecture studio AL_A, founded by Royal

**Superheated plasma is key to achieving fusion energy.
General Fusion's PI3 plasma injector is the largest,
most powerful fusion plasma injector in the world.**

General Fusion

Institute of British Architects Stirling Prize-winning architect Amanda Levete, to, as they put it, 'reimagine' their design.

General Fusion intend to use this demonstration plant to prove the economic and technical viability of MTF technology and so pave the way for a commercial pilot plant. The demonstration plant will be 70 per cent full-scale, and although it will not deliver power it will create fusion conditions at temperatures of over 100 million degrees Celsius. It will also operate at a rate of one plasma pulse per day – as opposed to the planned pulse every second in a commercial plant – allowing time for each cycle to be analysed and any improvements made to optimise the fusion process. Feedback on this demonstration plant from utility companies and other potential customers will be used by General Fusion to adjust their machine design to accommodate specific end-user needs.

Also combining a pulsed approach with magnetic confinement are Washington State-based Helion Energy, which was founded in 2013. Helion Energy are developing a pulsed, non-ignition fusion device for electricity generation that uses a fuel source of deuterium and helium-3. Their accelerator raises the temperature of the fuel to 100 million degrees Celsius, converting it into a plasma. Magnets are then used to confine the plasma into a Field Reversed Configuration (FRC), in which the magnetic fields invert a plasma on itself into a toroidal shape.

Helion Energy's fusion method relies on forming two of these FRC plasmas at opposite ends of a 12-metre-long (39-feet-long) accelerator. Magnets are then used to accelerate the two plasmas into one another so they smash together in the centre of the accelerator. The powerful magnetic field holding these combined plasmas in place creates enough

**A simplified diagram of the two plasmas inside the Helion
Energy accelerator, showing the path they take en route
to smashing together in the accelerator's centre.**
Helion Energy

compression to raise the temperature up to that required
for the deuterium and helium-3 ions to overcome their natural repulsion and for fusion reactions to occur. The result is
more energy being released by the fusion reaction than the
process is consuming and the plasma expands.

Rather than generating heat which turns water to steam
that drives turbines to create electricity, the Helion Energy
system has a more direct approach. The expanding plasma
exerts pushing forces on the magnetic field that confines it,
and thanks to Faraday's law, the changes this causes to the
magnetic field induce an electric current that can be captured
for use as power.

The fusion products are also captured since they include
helium-3 that can be reused for the next fusion pulse. Helion
Energy's patented closed-fuel cycle is an important part of
the process since helium-3 is so rare on Earth that there

has been talk of mining it from the Moon where it is much more abundant.

In 2020, the company finished construction of its sixth fusion prototype, called Trenta. Trenta has reached 104 million degrees Celsius (187 million degrees Fahrenheit), surpassing the minimum threshold of 100 million degrees Celsius at which enough fusion can occur for viable commercial energy generation via this method. While the company's long-term goal is to provide carbon-free electricity from fusion to the grid as soon as possible, at the time of writing, they are focusing on building their seventh fusion prototype named Polaris. The hope is that Polaris will demonstrate net electricity production from fusion in 2024.

Quick build times

Another private company setting the goal of 2024 for breakthrough results is Horne Technologies. Founded in 2008 by engineer Tanner Horne, Horne Technologies is taking a hybrid approach to fusion energy by combining four core technologies. The first is electrostatic heating. This involves electrostatic attraction from which energy gets transferred to the ions, thereby heating the plasma up to the temperatures required for fusion. The second main technology used is magnetic shielding and confinement of the plasma, while the third is a specific geometry for the magnetic coils. This enables the creation of a high plasma pressure compared with the pressure the magnetic field exerts on the plasma and is what plasma physicists refer to as 'high-beta' operating conditions. This is a desirable condition because it increases the gain of the system. Similarly cost-effective is the fourth core technology employed by Horne – rapid manufacturing techniques. It is Horne Technologies' unique hybrid confinement method that is allowing standard manufacturing technology to be applied to fusion for the first time.

According to Horne, all these factors help give his design the edge. 'Unlike the extremely expensive international projects I can iterate quickly and produce devices on budgets that most cannot imagine,' he describes.

In 2017, the company demonstrated an operational superconducting prototype reactor – the first hybrid design to be capable of continuous operation – and achieved their first plasma. At the time of writing, they are constructing a fusion-capable second-generation device with the aim to have it working in 2022. This will include various improvements over the first machine, including more advanced plasma diagnostics. Beyond this, Horne is already looking towards a third-generation reactor operating at full power by 2024. 'If all of the data [from the third-generation machine] looks good, after that [there] would be a net-energy demonstration device,' says Horne.

Unique configurations

As this and the previous chapter have revealed, the goal of providing smaller, cheaper fusion power plants than those which would result from using conventional tokamak technology is a driving force for many private fusion companies. This includes California-based TAE Technologies, which was founded in 1998 and has developed a prototype cylindrical colliding beam fusion reactor. This system, which is the fifth iteration since the company began its research and is named

'Norman' in memory of the late Norman Rostoker, who co-created TAE's fusion technology, uses hydrogen and boron as fuel, which are both non-radioactive. Firstly, at either end of the cylindrical reactor, hydrogen gas is heated high enough to form a ring of plasma. The two rings are then accelerated towards each other inside the reactor at faster than the speed of sound. This causes them to merge, resulting in a spinning plasma in the centre of the cylindrical device where fusion can take place. The plasma is then heated and sustained by particle accelerator beams and self-confining magnetic fields. The company calls this proprietary approach an 'advanced beam-driven Field Reversed Configuration'.

To assist with their experiments, TAE have been applying advanced computational methods, data science and machine learning that they developed in partnership with Google. As they strive to create a commercially viable fusion power solution for energy generation, TAE have also been developing new technologies that have applications in other industrial sectors. For example, because experiments with their fusion reactor required a far greater amount of power than their standard commercial grid feed supplied, the company needed to develop more efficient energy storage solutions as well as new methods of power delivery. This same power management technology could also be applied in other areas, such as electric vehicle charging stations – enabling faster charging – for residential energy storage, and in computer data centres. The latter are notoriously power-hungry and are growing in size all the time as the demand for streaming movies and music rises and as many of us increase the amount of content we upload to cloud storage systems. By 2006, servers and data centres in the United States were already accounting for 1.5 per cent of total US electricity consumption.

The global increase in electricity consumption is, as Chapter 8 will detail, a major driver for fusion research. But as much as the world could do with rapid progress when it comes to fusion energy, there are some constraints. As we are about to see, no matter what technological approach is being taken, the engineering needed to create experimental fusion reactors is not only demanding to design but is also stretching the limits of worldwide manufacturing capacity and capability.

AN ENGINEERING FEAT 5

No net power

Creating a nuclear fusion reactor is by anyone's standards a major engineering feat. The architectural and engineering challenges posed when housing a reactor are numerous and complex.

It certainly doesn't help that for fusion reactions to occur on Earth, the temperatures have to be much higher than the 15 million degrees Celsius (27 million degrees Fahrenheit) in the Sun. Plus, as we saw in Chapters 3 and 4, without the enormous gravitational forces in the Sun compressing the fuel and facilitating fusion we have to use other methods of confinement for plasma on Earth to reach the right conditions to enable the process.

In the Sun, the amount of fusion power produced per unit volume, in other words the fusion power density, is, perhaps surprisingly, relatively low. But because the Sun is so vast in size it still produces a lot of power overall.

Fortunately, our fusion reactors on Earth have a much higher power density, thereby enabling the plasma volume to be relatively small but still give out a usable amount of fusion power. This means they can be tiny in scale compared with the Sun while still producing a lot of power, albeit only by using specific combinations of fusion fuel. These include the deuterium–tritium option due for use by ITER.

But even when the incredibly high temperatures have been reached and the plasma has been confined, the enormous challenge of finally achieving net energy gain from the fusion reaction itself remains. This is something which is proving stubbornly elusive so far. Fusion technology is of no use as a power plant if you put in more energy than you get out, and at the time of writing, no one has yet cracked continuous operation with net energy gain. Until this is achieved, working reactors are some way off.

We saw earlier that vast amounts of energy are released in nuclear fusion reactions. So, from a physics perspective, what is the problem with creating net energy in a fusion reactor?

Fitting the criteria

In a nutshell, you need to satisfy the Lawson criterion for the net release of energy. This was established by British physicist and engineer John David Lawson (1923–2008), who worked on fusion energy research. In 1957, Lawson showed mathematically the minimum conditions for a fusion reaction to release more energy than is being put in. This means achieving a value greater than the following three quantities multiplied together: first, a measure of how well the confinement system holds on to the heat – in the same way that a Thermos® flask will keep your tea or coffee warmer for longer than just putting it in a mug. This is known as the energy confinement time. Second, the plasma temperature. Third, the plasma density; in other words the number of particles within a given volume of the plasma.

It turns out that different approaches to fusion have different challenges when it comes to satisfying the Lawson criterion. This is because the different approaches to the physics and engineering create a specific set of conditions. For magnetic fusion energy approaches, including tokamaks and stellarators, the plasma inherently has a low density but a high energy confinement time. Inertial fusion systems, such as that at the National Ignition Facility in the United States, are the opposite as they have a high plasma density but a low energy confinement time. Meanwhile, hybrid magneto-inertial systems have medium values for the plasma density and energy confinement time.

Since reaching the minimum number that must be exceeded for net energy production means multiplying these quantities and the plasma temperature together, it is clear that each approach presents different physics and engineering challenges.

That said, there are some common problems for all fusion approaches, including how to confine the plasma within the reactor chamber. As we have explored, the plasma inside a fusion reactor is at a staggeringly high temperature of millions of degrees Celsius, and would simply melt any surfaces that it came into direct contact with. So, methods are needed to keep the incredibly hot plasma in a fusion reactor safely away from the reactor walls.

Reining it in

For most magnetic confinement systems such as tokamaks and stellarators this is achieved via extremely powerful electromagnets. Many of us will use devices containing electromagnets on a daily basis. You will find them in electric fans, microphones, audio speakers and doorbells. The beauty of electromagnets is that they are not permanent magnets. They are made from a coil of wire, known as a solenoid, that an electric current runs through. Their magnetism is essentially switched on and off via the current because it is only when electricity flows through the coils that a magnetic field is created.

This means electromagnets can be used to operate switches and move parts by attracting components while they are magnetised which then fall back away once the magnet is switched off. In the case of a doorbell, pressing the button completes an electric circuit which activates the electromagnet. This then attracts a metal arm with a striker that hits the gong causing a sound. As it moves, the metal arm breaks the circuit turning the electromagnet off. If the visitor is still pressing the button on the doorbell, as the arm returns to its starting position it reconnects the circuit, allowing current to flow again and magnetise the electromagnet, which leads to another ding on the gong.

The electromagnets required for most magnetically confined fusion reactors are, as you might expect, in another league compared with your doorbell's electromagnet when it comes to power. They need to be able to create a magnetic field strong enough to sufficiently push on the plasma to keep it away from the reactor walls. Creating such magnets has been one of the major technological challenges for ITER, and other fusion energy projects using magnetic confinement via external magnets. And this challenge simply could not have been met without using a type of material discovered in 1911 by the Dutch physicist Heike Kamerlingh Onnes – a superconductor.

Conducting proceedings

We are familiar with electrical conductors, such as copper, aluminium and gold. These materials all conduct electricity well. Conversely, insulators like rubber are so poor at conducting electricity that they can be used to cover electrical wires and encase electrical equipment, keeping us safe from shocks and burns. While the electrical wires inside the insulation are usually made of copper, which conducts very well, these wires do still offer up some resistance to the flow of an electric current. That means some power is lost as the electricity moves through the cable. But the superconductors discovered by Onnes provide almost no resistance to an electric current at all.

Superconductors can be thought of as the superheroes in the world of electrical conductors. The conductivity of a superconductor is about 100,000 million, million times greater than the conductivity of a typical metal. Astonishingly, experiments have suggested that if a direct current was set up in a continuous loop of superconducting wire that this current would continue flowing round the loop for 1,000 years. The superconducting wire would offer up a very small amount of resistance to the current flow, but not enough to reduce its conductivity by much during a human lifetime.

While there is no need for a tokamak or stellarator fusion reactor electromagnet to be switched on for 1,000 years, superconductors are needed to create the very powerful magnets required. This is because the larger the electric current flowing through an electromagnet, the more powerful the magnetic field produced. Superconductors are therefore the only choice for magnetically confined fusion reactor electromagnets.

Confined space

To confine a plasma inside a tokamak, two different shapes of magnetic field are required – a toroidal magnetic field and a poloidal magnetic field. The toroidal field has a series of separate magnetic field lines, which loop around the reactor vessel a bit like curtain rings on a pole. If you imagine your curtain pole is flexible enough to bend it round into a circle, you will have a pretty good mental image of how the toroidal field looks in a tokamak, with the pole representing the reactor chamber. To create each of the individual curtain-ring-shaped magnetic fields, the tokamak needs a separate toroidal field coil. Ten field coils would therefore produce ten separate toroidal magnetic fields, which would sit like ten curtain rings spaced apart from one another around your circular pole.

By contrast, the poloidal field runs around the circumference of the doughnut-shaped reactor chamber. So, using the analogy above, the poloidal field would follow the shape of your circular curtain pole.

The superconducting magnet system is one of the key design features of the Korea Superconducting Tokamak

Advanced Research project. To contain the plasma in the KSTAR reactor, there is a magnet system made up from sixteen toroidal field coils and fourteen poloidal field coils.

Meanwhile, in ITER, there will be a central solenoid, eighteen 'D'-shaped vertical magnets – each weighing in at 360 tonnes (397 US tons) and measuring 9 by 17 metres (29.5 by 56 feet) – that conform to the toroidal shape of the magnet vessel, and six horizontal ring-shaped poloidal magnets. The latter are arranged a bit like tableware in a plate rack, with the largest having a 24-metre (79-feet) diameter and the heaviest a scales-busting weight of 400 tonnes (441 US tons).

Used in combination, these magnets create an invisible magnetic cage that confines the plasma inside the walls of the vacuum vessel. You need this because you are taking the temperature of the plasma ten times hotter than the core of the Sun, and since no material on Earth can withstand that temperature, the magnetic cage keeps the plasma away from the walls.

The ITER magnets are made from niobium–tin or niobium–titanium and become superconducting once cooled – via liquid helium – below –269 degrees Celsius (–452 degrees Fahrenheit). Despite the specialist materials they are made from, and the fact they need to be cooled to an incredibly low temperature, these superconducting magnets give more than just the necessary power. They are also cheaper to operate than conventional electromagnets and – by their very nature – consume less electricity.

The scale of these magnets is difficult to comprehend. All of the ITER magnets combined will weigh in at 10,000 tonnes (11,023 US tons), and more than 100,000 kilometres (62,000 miles) of superconducting strands are needed to

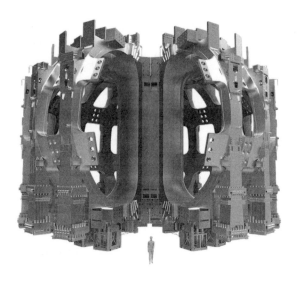

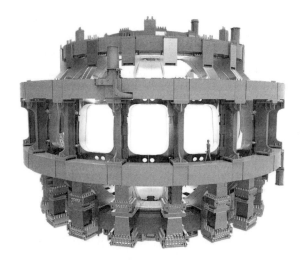

**An overview of ITER's toroidal field system (top)
and poloidal field system (bottom).**

ITER Organization

make the toroidal coils and central solenoid. (The superconducting strands are twisted together with copper and encased in a steel jacket to create high performance conducting cables.)

The creation of these magnets is pushing the boundaries of what is possible from a manufacturing perspective. Not least because of how technically challenging it is to produce niobium–tin superconducting strands.

Because these superconducting magnets are so large, they require components known as 'magnet feeders' to work. These magnet feeders are also on a big scale, with each weighing 6.6 tonnes (7.3 US tons) and 10 metres (33 feet) in length. They 'feed' the superconducting magnets with electrical power and ultra-low temperature fluids, as well as linking them via instrumentation cables to the world outside the reactor.

To compensate for any slight deviations in manufacturing tolerances for the other superconducting magnets, ITER will also have eighteen superconducting correction coils arranged in groups of six outside the reactor chamber. These coils will be able to tweak the shape of the magnetic field confining the ITER plasma.

A stable relationship

As we have just seen, plasmas inside tokamak fusion reactors require a lot of magnets to keep them stable. This is because they behave a bit like a bath sponge when you squeeze it between your hands. While you are squeezing you are applying a compressive force to the sponge. But as soon as you let go, the sponge will spring back into its original shape.

The plasma inside a fusion reactor is similar in that the forces from the pressure it is under cause it to want to jump into another shape. This shape would have a lower energy compared with the shape the magnetic fields are forcing it into because physics dictates that any systems – from large plasmas down to tiny atoms – will naturally move into the lowest possible energy state unless something is preventing them from doing so. This means that the magnetic fields of a tokamak or stellarator require constant adjustment since the plasma is always fighting against being pushed into a specific shape.

The inherent overall instability of a confined plasma also causes relatively small, localised instabilities at its edges. These need to be carefully controlled to avoid adversely affecting the fusion performance or damaging the reactor wall components.

But even with these complex magnet systems keeping the plasma from touching the reactor walls, the wall materials will still be subject to incredibly high temperatures, which raises the question of how to figure out what materials to make these plasma-facing components from?

Material benefits

This is by no means a straightforward task, as these materials will be subject to the most extreme conditions ever faced by anything on the planet.

For example, within the ITER reactor, the armour covering the inside of the reactor walls and the divertor (through which spent fuel flows out and new fuel is injected in) will come closest to a plasma that is ten times hotter than the

Sun. Not only that, but both these components will also have to operate under an ultra-high vacuum with a pressure 10 billion times lower than atmospheric pressure. So, the ITER team began their quest by compiling a list of likely candidate materials.

Firstly, these materials needed to absorb up to an astonishing several million watts per square metre of heat energy density from the nearby plasma. At its hottest, the heat they would face would be over 10,000 times more than we would experience lounging on a Mediterranean beach in the midday sun.

To cope with such high heat energy, the materials chosen for both the armour and the divertor must have high thermal conductivities. Thermal conductivity is defined as the rate at which heat passes through a small area of a body. So, it is a measure of how well a particular material can conduct heat. We know from everyday experience that a metal spoon left in a hot cup of tea will get hot itself, while thick cotton oven gloves will prevent us from burning our hands when we get dishes out of the oven. This is because metal has a higher thermal conductivity than cotton.

Of course, the high thermal conductivities of metals like silver and copper have been known about for years. So, when it came to choosing materials for use in ITER, there was a wide range of high thermal conductivity materials out there. But there was an additional problem that these plasma-facing materials had to cope with: contamination.

The worry was not whether the materials chosen would become contaminated, but the opposite. Could the plasma-facing materials themselves pollute the plasma and interfere with the fusion reaction, thereby destroying the very thing that they were trying to protect?

The ITER scientists knew that there was no way of avoiding some of the incredibly hot plasma particles hitting the surface of the plasma-facing materials and knocking atoms out of that material. These freed-up atoms would then go into the plasma, contaminating it. This type of surface erosion is in essence an occupational hazard of creating fusion on Earth. So, to keep the plasma as pure as possible, the ITER team began a process of elimination for choosing the best option for the plasma-facing materials.

They knew that the more protons in the nucleus of the chosen material, the more harmful it would be to the plasma. This is because when eroded atoms from plasma-facing surfaces of reactor components go into the plasma they bump into the plasma particles. The changes that occur as a result of these collisions lead to the plasma cooling down, which is not something you want to happen because cooling the plasma will adversely affect the fusion reaction. Since the more protons there are in the contaminant atoms the more severe these adverse effects are, it is vital to choose plasma-facing materials with the lowest number of protons possible.

But this was only part of the story. It is also the case that the greater the number of protons and neutrons in the nucleus, and so the higher the mass of the atom, the harder it is for an atom of that material to be eroded from a plasma-facing surface.

By looking at the Periodic Table, which lists the elements in order of the number of protons in their nucleus, the ITER scientists started to narrow down their materials choice. Since both hydrogen and helium are gases at the temperatures required for the reactor, and the next element lithium is a liquid, they could not be used. But beryllium, with just four protons in its nucleus, is a solid metal. Beryllium also has a high thermal conductivity. Better still for ITER, if the contamination from the beryllium could stay at just one beryllium atomic nucleus for every hundred nuclei of deuterium and tritium fuel, the plasma would not be adversely affected. As it happens, the operating conditions for ITER should not cause any more contamination than that. Therefore, even though beryllium is light and so can be eroded relatively easily, it was chosen for the 8 to 10-millimetre-thick armour for the reactor walls. (This is not the only recent occasion beryllium has proved attractive to scientists. Beryllium's high thermal conductivity was part of the reason it was chosen to make the mirrors in the recently launched James Webb Space Telescope, as it helps them withstand the heat they receive from the Sun.)

The ITER divertor needed a different material because the spent plasma fuel directly touches this component. This means it has to withstand the highest temperatures and amount of particles of any part of the ITER machine. Tungsten was the material chosen, partly because it has a melting point of 3,422 degrees Celsius (6,192 degrees Fahrenheit).

Although the plasma cannot tolerate anywhere near as much contamination from tungsten – anything over one tungsten nucleus per 100,000 nuclei of deuterium and tritium fuel is enough to destroy the plasma – it does have the advantage that it is much trickier for the plasma to remove atoms from tungsten's surface than it is for it to erode the much lighter beryllium. Plus, the divertor is over 1 metre away from the plasma, making it harder for any eroded tungsten atoms to reach the plasma compared with beryllium atoms, which only have to travel a few centimetres to cause havoc in the plasma. Tungsten therefore works well as a materials choice for the divertor.

Fortunately, tungsten is a common metal. But some of the other materials used in ITER and other fusion reactors, including beryllium, are not so readily available.

Too hot to handle

It turns out that if you're a fusion physicist, diamonds are not your best friend. Emeralds are. So too are aquamarines. That's because beryllium is a by-product of the processing of both gemstones, which are part of the beryl family and contain beryllium.

Beryllium is so useful to the nuclear industry not just for the reasons outlined in the previous section but because it is the most effective multiplier of neutrons that exists. Since you need a lot of neutrons to get atomic weapons to explode, the first nuclear use for beryllium was actually in the warheads of atomic bombs. This use led to the United States starting up surface mining for the beryllium-containing ore bertrandite, which is more plentiful in the United States than emeralds or aquamarines and provides a domestic source of beryllium.

But beryllium also has a range of peaceful nuclear roles as it is used for research into how different materials perform when they are irradiated in nuclear environments, and in the production of medical radioisotopes. As we have learned, beryllium is also an excellent choice for some components in fusion reactors; however, it is not without its problems. It is a toxic material and is also highly reactive, second only to magnesium in terms of its reactivity. In fact, its high reactivity is why it does not naturally occur in elemental form and needs to be refined from the mined mineral bertrandite. Along with its toxicity, beryllium's reactivity also makes it very tricky to handle.

Safety first

But chemical hazards from materials such as beryllium are not, of course, the only safety concern when it comes to fusion reactor design. Although fusion does not itself produce radioactivity, as mentioned earlier, the neutrons created in the process can make some of the materials a fusion reactor is constructed from radioactive. These materials can also be transformed into other materials by transmutation – for example, neutron irradiation can turn gold into mercury – or their mechanical properties, such as their strength and hardness, can be altered.

The neutrons can cause so much havoc because they have no electric charge, so the magnetic field confining the plasma has no effect on them. This means they are free to move in all directions after being emitted in the fusion reaction, which enables the neutrons to hit into and so interact with the nuclei of any of the surrounding reactor materials.

Neutrons can interact with the reactor material nuclei in several different ways, the probability of each depending on the energy of the neutrons and the properties of the nuclei they are colliding with. One possibility is that neutrons bounce back off the nuclei in the reactor material, transferring some of the energy from their motion to the nuclei. This heats up the material and potentially damages it if atoms then move around within its structure. By contrast, another potential interaction results in the neutrons being absorbed by the material they hit and effectively stopped in their tracks – such materials are known as neutron absorbers. Neutrons can also be absorbed in such a way that a proton or other type of radiation is emitted and causes transmutation of the reactor material. Yet another prospect is that neutrons may collide with the nuclei of reactor materials in a manner that leaves the nuclei in a higher energy state, which can then undergo radioactive decay.

In itself, a material becoming radioactive does not necessarily mean it can no longer do the job it was designed for. So, in the sense of ensuring the materials a reactor is made from are still up to the task, it is not inevitable that radiation will cause an issue. Plus, there are mitigations you can put in place to quell the effects of the neutrons. One is to strictly control the number and nature of impurities in the materials that will be receiving a neutron dose as this reduces the number of interactions that lead to radioactivity. But even though this helps, there is still a certain amount of radioactivity that is created within a fusion reactor. This, not surprisingly, needs managing carefully so that it does not cause any safety issues for workers or the surrounding environment.

The main mitigation strategy against this radiation is to use standard nuclear industry shielding materials in and around reactors. These include concrete and steel with added boron, known as 'borated steel'. Both are capable of stopping both gamma radiation, which is produced when fast-moving ions in the plasma react with fuel ions or contaminants, and neutrons escaping from reactors into the outside world. One of the reasons concrete is chosen is because its structural properties are well understood. For many scenarios, this practical consideration outweighs the fact that concrete is not actually the most efficient material for radiation shielding.

Any shielding must of course be sufficient to protect workers and members of the public from the radiation inside the reactor. Consequently, in ITER there are locations requiring a higher shielding efficiency than standard concrete can offer. These areas are typically where space constraints mean the concrete cannot be as thick as you might ideally wish to choose. In these instances, a heavy concrete with a denser aggregate and higher density is used. Further shielding efficiency can be achieved by using 'borated concrete' – made by mixing boric acid powder into concrete – which is better at absorbing neutrons. A low-sodium version of concrete is also preferred for shielding because sodium, which can be naturally present in the concrete's aggregate or in cement, depending where it is mined, can become intensely radioactive for a short period of time.

To shield ITER's superconducting coils, two types of borated steel are being used. One has 4 per cent boron content and the other has 7 per cent boron. Because boron makes steel brittle, the 7 per cent boron version is only used where it is unavoidable because that specific location leaves little room for the necessary shielding. One of these places is between the inner and outer walls of the vacuum vessel where 40-millimetre-thick borated steel plates containing up to 7 per cent boron need to be sited at the most critical places. They are to provide shielding for the superconducting magnets and also help to control the amount of induced radioactivity in external structures.

In locations where there is more space available, and so a thicker layer of steel can be used, the 4 per cent boron version will be employed because it has better tensile properties.

In other words, it is stronger when it is under tension, which can be a critical property to have in certain locations.

In order to work out where, and how much, shielding is required, fusion physicists and engineers need a good understanding of radiation conditions throughout their reactor building. They also need to know what effects any neutrons might have on the components of their machine. Although some experimental studies can be carried out, in the main, these challenges need to be investigated via modelling. But unlike the models we explored in Chapter 1, created by Ernest Rutherford and his contemporaries in the late 19th and early 20th centuries, this modelling involves computers. And in the case of estimating the effects of neutrons by looking at how likely they are to interact with atoms and ions and induce radioactivity, it requires the solving of thousands upon thousands of mathematical equations – a feat which can only be performed by incredibly powerful supercomputers.

While some of the calculations can be based on decades of past experience from either current nuclear (fission) power stations or from experimental fusion reactors, each fusion project will have its own unique set of radiation shielding challenges so will require different modelling.

One of the most-used mathematical techniques for looking at the effects of neutrons is known as Monte Carlo modelling. Instead of creating an absolute answer to a problem being posed, this technique involves carrying out a series of calculations tracking billions of individual neutrons within a set of specified parameters to come up with the probability of the event you are studying occurring. Monte Carlo modelling can be carried out by hand as well as by computers. But the more calculations made, the more accurate a result you get, so computers are essential for studies related to fusion and the wider nuclear industry. Since Monte Carlo methods are also used in other areas, including the life sciences and weather forecasting, the desire for better accuracy and more sophisticated simulation has helped drive advances in computer technology – as has the desire for a more realistic computer gaming experience for millions of gamers through the world.

In the case of looking at neutrons travelling through the ITER machine and the surrounding building, for example, a computer model is firstly constructed that contains all the details of the reactor components. Then the Monte Carlo method makes a random choice as to what interactions a given neutron will make with the materials the machine is made from. By repeating this calculation over and over and over, the result, showing average behaviour of a neutron in the ITER machine, becomes so accurate that reliable design choices can be made for the materials used and the level of shielding.

To make the vast numbers of calculations required for accurate results you need an equally vast amount of computer processing power, making this just one of the areas in which the latest designs of fusion reactors are being assisted by advances in computer technology.

Digital revolution

It turns out that when it comes to fusion reactors, it is not just the physical architecture of the site that is impressive. The computing architecture that has helped build the site is also spectacular. For example, in order to model the neutron behaviour, and so work out the shielding required for ITER, a supercomputer called MareNostrum at the Barcelona Supercomputing Center was used to carry out the calculations.

The latest version of this beast of a machine has fourteen petabytes (PB) of storage, or 14 million gigabytes (GB), which is equivalent to the storage found in 14,000 laptops, assuming they each have a 1 TB (1,000 GB) hard drive. Users book time on the supercomputer, which is connected to European research centres and universities via two networks. These are the RedIRIS network that connects over 300 Spanish academic and scientific institutions and the Europe-wide GÉANT network that connects more than 10,000 research and education institutions. When operating at peak power, MareNostrum can perform in excess of 11,000 trillion operations per second.

Using MareNostrum is a far cry from the sorts of computers available for fusion research in the early 1980s, when Michael Loughlin, nuclear analysis and shielding coordinator for ITER, entered the field.

'That was the era when home computing became popular. I remember colleagues stacking many of these cheap computers together and running calculations in parallel,' he recalls. This 'cluster' of computers enabled calculations not previously feasible, and soon technology had advanced to the point where a single chip was the equivalent of having several connected computers working together.

As computers became ever more powerful, Loughlin was able to run Monte Carlo calculations over a course of several days that would have taken several years to complete with the previous generation of computers.

But even with the incredible computing power available now, there is only so far that engineers and scientists can get with computer modelling. In ITER, for example, this will be the first time superconducting magnets have been operated in a superconducting state in the presence of so many neutrons and gamma rays. The ITER scientists are already gearing themselves up for potentially seeing some unexpected effects, such as either more heating or less heating than predicted in the superconducting magnets. As a result, the modelling of the neutron behaviour will be an ongoing part of the work, not least because the reactor will remain operational for many years and will need routine maintenance throughout this time.

Efforts must therefore be made to minimise any dose that any workers might be subjected to in the course of any maintenance procedures. In the early stages of ITER operation, when the power levels are low, the scientists will check their calculations to make sure that they match what they are observing in the real world situation. If they find any discrepancies, additional shielding can be added.

Of course, modelling neutron interactions is not the only area in which advances in computer technology can assist with the development of fusion power. Computer Aided Design has made the drafting of designs of fusion machine components easier, while plasmas are analysed and controlled using state-of-the-art computer components.

For example, a team from the University of Washington (UW) in collaboration with private company CTFusion are using a graphics processing unit (GPU) to run the system that controls the formation of the plasma in the UW prototype fusion reactor. (More generally, these types of computing operations would be carried out via software running on a central processing unit (CPU), so using a GPU instead makes for a unique type of control system.) They chose an

Nvidia P40 Tesla GPU – which has the same architecture as a gaming graphics card but is more powerful – because it has the speed and precision needed to react to the constantly evolving shape changes that happen at very high speed in their plasmas.

So far, the researchers have reached temperatures of around 1 million degrees Celsius (1.8 million degrees Fahrenheit) in their reactor, which although much lower than the temperature needed for fusion is hot enough to study how the plasma is behaving under its magnetic confinement. This confinement is courtesy of magnetic fields generated within the plasma itself rather than from external magnets. The fast-acting control system run by the high-speed GPU is able to fine-tune these magnetic fields so that the plasmas, which can only be maintained for a few thousandths of a second, remain under control.

Everything but the kitchen sink

Although most aspects of fusion reactor design are complex in one sense or another, it is not the case that all the materials used are equally high-tech, or that all components need their use planned or their operation carried out by huge computers. In ITER for instance, carbon steel is used for some of the big, heavy doors that are a reasonable distance away from the reactor chamber. Another familiar material that is used in ITER is stainless steel. Albeit that this stainless steel is not the same type used to make our knives and forks, or our kitchen bin.

Stainless steel comes in a range of different variants, each with slightly different constituents and properties, which are selected depending on its end use. In ITER, they use several different steels. For example, a low-nitrogen and low-carbon stainless steel known as SS316 is used in locations where not only the strength and magnetic properties offered by steel are required, but where the metal will also need to cope with being subjected to large amounts of radiation.

Procurement challenges

But even though some standard materials are used in fusion reactors, there are a lot of components that do require specialist materials or manufacturing facilities. For instance, we discovered earlier that fusion reactors require superconducting wire to make their powerful electromagnets. Not only are these wires extremely high spec, requiring specialist manufacturing techniques, but also an awful lot of them are needed. For example, ITER will use 10,000 tonnes (11,023 US tons) of superconducting magnets to create a field that will stop the plasma from coming into contact with its reactor walls.

Manufacturing the vast quantities required of some of these components is a challenge in itself. When ITER procurement began in 2007, the worldwide production of the superconducting strands they need was about 20 tonnes (22 US tons) per year. To produce the 200,000 kilometres (125,000 miles) of superconducting wire required for the ITER reactor meant ramping up global production of the strands to more than 100 tonnes (110 US tons) per year. This was while maintaining the same performance and quality from their ten suppliers who each had differing capabilities.

Interestingly, there have already been some knock-on benefits of this production increase. Not only are ITER-quality superconducting strands used in an upgrade to the Large Hadron Collider (LHC) at CERN, they are also being used in high-performance medical scanners.

At the National Ignition Facility (NIF) in the United States, one of the most important components for their fusion experiments is the specialist laser glass required to amplify laser pulses up to the incredibly high powers and energies required to create fusion conditions. When they were building their latest laser system in 2005, the NIF needed over 3,000 42-kilogram plates of this glass, which contains the rare earth element neodymium. Each 3.4-centimetre-thick plate measures 46 centimetres by 81 centimetres, and if all the plates were placed end-to-end, the resulting ribbon of glass would stretch for over 2.4 kilometres (1.5 miles). But at that time, producing this amount of glass within the time frame needed by the NIF was simply not feasible. It meant a new production process had to be developed by suppliers Hoya Corporation, USA, and SCHOTT North America.

This new process involved melting the raw materials for the glass into a single continuously flowing strip of molten glass, which, once cooled, could be cut into pieces and polished ready for use in the NIF. Compared with previous processes, the new manufacturing method enabled glass of better optical quality to be made twenty times faster and five times cheaper in price.

Sourcing raw materials can be equally challenging. The beryllium required for future fusion reactors, for instance, will also require a huge ramping up of production. This is because while ITER will use around 14–15 tonnes (15–17 US tons) of beryllium in total, a working commercial fusion power plant would require around 200–300 tonnes (220–331 US tons) of the material. Although beryllium is rare in the sense that not much of it is mined and processed per year, fortunately for the fusion industry, it is not scarce. It should be quite possible to excavate more of it out of the ground.

Customers will also need to dig deep in their pockets as beryllium doesn't come cheap. As a bulk, engineered material, it can be 100–1,000 times the price of aluminium, and in ultra-thin foil or in ultra-high-purity forms, it can be more expensive than gold in price per gramme. Each beryllium atom forms such a strong chemical bond with its neighbouring atoms that it is very tricky to get out of the mineral it lives in. That difficulty translates to a complex refining process which pushes the price up compared with elements that are easier to obtain.

Even once all of the construction and materials challenges have been solved for reactor designs, further hurdles still need to be overcome. In particular, the safe handling and transporting of fusion fuel and dealing with both spent fuel and other radioactive waste. Fortunately, fusion is inherently safer than nuclear fission. But as we are about to see, operating and maintaining a fusion reactor safely still requires some specialist solutions.

TIDYING UP AFTER 6

Catastrophic consequences

On 11 March 2011, Japan was struck by a powerful, magnitude nine earthquake – the fourth most powerful ever recorded. In addition to the direct damage and fatalities it caused, the earthquake triggered a tsunami so large that it flooded up to 10 kilometres (6 miles) inland, wreaking devastation to whole towns and drowning thousands. Over 18,000 people are estimated to have lost their lives as a result of this earthquake and tsunami. In addition, almost 400,000 buildings were either destroyed or damaged beyond repair, with a further 750,000 buildings partially destroyed. Sadly, over the next few days, the disaster continued to unfold.

Directly in the path of the tsunami, on the north-eastern coast of the largest of Japan's main islands, Honshu, was the Fukushima Daiichi Nuclear Power Plant. As with all current nuclear power stations in the world, this plant relied on nuclear fission to create electricity. As soon as the earthquake was detected, automatic systems turned off the Fukushima

Daiichi reactors. But the deluge of water from the tsunami caused the emergency back-up generators at the plant to fail, which in turn led to the cooling systems failing.

Pumping a coolant (commonly water, but other liquids or gases, including air, carbon dioxide, helium and liquid sodium, can be used depending on the reactor design) through the core of a fission reactor removes the heat. This process not only captures the heat that creates the steam to drive the turbines that produce electricity, the cooling system is also essential for avoiding so-called 'meltdown' of the reactor core. In a meltdown, nuclear fuel, and potentially its container too, become so hot that they turn into liquid, causing the reactor to collapse. So, once the cooling systems failed at Fukushima, efforts were quickly made to try to restore power to them.

But it was too late. Three of the four reactors at the plant went into meltdown. There was also a series of explosions. The net result was a release of radioactive material into the surroundings that meant more than 100,000 people were evacuated to avoid radiation exposure, and an exclusion zone was created around the nuclear plant. It was the worst nuclear accident since the Chernobyl disaster of 1986, in which flaws in both the Soviet reactor design and the safety culture led to a series of explosions and fires that released radioactive material into the air, which went on to contaminate many areas of Europe. In addition to what the best estimates suggest is fifteen later deaths from cancer, the initial explosion killed two plant workers, while a further 28 people – comprising their colleagues and six of the firefighters first on the scene – died from acute radiation syndrome within a few weeks of the disaster. Work to clean up the site is still ongoing.

At Fukushima, the Japanese government has since spent tens of billions of dollars on the ongoing clean-up operation there, although most of the exclusion zone remains uninhabited by humans to this day.

On the safe side

Clearly, radioactive contamination and exclusion zones that have to remain in place for years are far from ideal, and there are some similarities between the fission and fusion processes. Both are nuclear reactions. Both also involve the famous equation $E = mc^2$, which as we saw in Chapter 1 explains how the reactions convert a tiny bit of mass into a large amount of energy. But when it comes to the physics, and also to the engineering involved in building a reactor, fission and fusion could not be more different. Therefore, do we need to be worried that a Fukushima-type incident could occur at a fusion energy plant?

The short answer is no. Both the reactor design and the fusion reaction itself make that scenario completely impossible. This is for several reasons.

Firstly, for a fusion reaction to be sustained there must be a continuous fuel supply. Consequently, if there is a problem with a fusion reactor, you simply stop adding more fuel. Without a constant supply of new fuel, the fusion reaction ceases of its own accord. (A point we will look at in more detail shortly.)

The quantities of fuel required by the different types of reactor are also very different. A fission power plant uses several hundred tonnes of uranium or plutonium during operation, whereas, in a fusion plant the most fuel you would have in the reactor at any given time would be 2 to 3 grammes of deuterium and tritium, which means that there is a lot less fuel to look after at a fusion reactor site.

Looking at the two different reactions in more detail, the nuclear fission reaction in today's nuclear power stations can be thought of as being a bit like passing on a virus. Say you start with one infected person, who then passes the virus on to two others. If those two other people then each pass the virus on to two more people, before you know it, that's seven people who have gone down with the virus. If that same process continues, the total number of people infected gets larger and larger very quickly.

It's the same for nuclear fission. As one nucleus is split into two, neutrons fly out and hit into other nuclei causing them to split into two. The neutrons produced from those reactions then cause further fission, and very quickly the situation multiplies and creates what nuclear scientists call a chain reaction.

In fact, in a fission reaction it is a uranium nucleus (generally) that is split apart with an incoming neutron. The result is smaller nuclei and the release of a series of neutrons which, once they have slowed down sufficiently, then trigger more fission. For a viable power station you need a lot of neutrons available. This is because alongside creating the chain reaction, it is neutrons transferring their energy to the water circulating around in the reactor walls which in turn creates the steam that drives the electricity turbines.

But of course, the last thing you want is the chain reaction multiplying indefinitely. A lot of the technological features in the current generation of nuclear power plants are, therefore, designed to stop the fission reaction from running out of control. If this happened, we would be talking about a major disaster and mass evacuations of surrounding areas.

To avoid this, that chain reaction must be kept at a 'critical' level. Below that level of criticality and the fission reaction doesn't work. Above that level you can get a runaway. Hence, you must moderate the reaction. It is worth noting that some nuclear terminology can sound much scarier than it is. For example, when a reactor 'goes critical' that simply means it is in a working state where there are enough reactions to maintain a sustained, but controlled, chain reaction. Essentially being 'critical' means a fission nuclear power station has been switched on.

After it has been turned on, the rate of the fission reaction is kept within the required limits using control rods. Control rods are usually made from materials such as boron, cadmium or hafnium, which are very good at capturing the neutrons released in the fission process. With the neutrons absorbed, they cannot go on to create new fission reactions. So, to turn off power generation, the control rods are inserted into the nuclear reactor. They act like an 'off' switch.

If anything prevents you from operating that off switch then that is when you could run into problems. When a fission reaction is uncontrolled, the build-up of heat that results must be continuously cooled down. In the event of a broken cooling system, you can get a nuclear accident of the type seen at Chernobyl or Fukushima.

By contrast, although it is much more powerful than a fission reaction, the fusion reaction is not a chain reaction. This in itself makes fusion safer than fission.

Also, because for a fusion reaction to proceed you need to supply an extraordinarily high level of heat, if you turn off the heating mechanism at any stage before the plasma ignites, the fusion reaction ceases. Once the plasma reaches ignition – the point at which all of the heat needed to keep the plasma hot enough for fusion is coming from the fusion reaction itself – the situation changes. At this point, you cut back on the external heating that was required to get the plasma up to its ignition point as it is no longer necessary. The only way to keep the fusion reaction going after this point is to add more fuel. If there is a problem with the reactor you just stop feeding it with more fuel. No fuel, no fusion.

Another factor which makes fusion safer is the design of a fusion reactor compared with a fission reactor. The big difference is that if a fusion reactor's cooling system failed completely during operations, the vacuum vessel that contains the nuclear fuel simply could not get hot enough to melt the materials it is made from. In fact, a fusion reactor's cooling system is not considered critical to safety for two reasons. Firstly, because of the difference in fuel mass: as we explored earlier, the amount of fuel in a fusion reactor is a few grammes of hydrogen (deuterium and tritium), whereas a fission reactor normally has about 200 tonnes of uranium in the core at any given time. Secondly, because fusion is not a chain reaction.

Consequently, in complete contrast to a fission reactor, a meltdown cannot occur for a fusion plant. This difference in the physics of fission and fusion is why plasma physicists refer to the fusion reaction as 'inherently safe'.

But even though meltdowns are thankfully impossible for fusion reactors, what radioactive waste products are produced from the fusion process under normal operations? Also, how can radioactive components be dealt with once they reach the end of their working lives?

To answer these questions, we first need to understand

exactly what it means for something to be radioactive and how long it remains a risk.

Keeping active

Radioactivity was first detected by the French physicist Henri Becquerel in 1896. It occurs when nuclei disintegrate, releasing either energy or particles, or sometimes both. This process is known as radioactive decay. As discovered by Ernest Rutherford, the physicist we met in Chapter 1, there are three types of radiation that can be emitted from a disintegrating nucleus during decay: alpha particles, beta particles and gamma rays.

An alpha particle is composed of two protons and two neutrons, so is the same as a helium nucleus. In alpha decay, the loss of these four nucleons means that the nucleon number of the decaying nucleus decreases by four, therefore forming a different type of atom. The decay of uranium-238 into thorium-234 is an example of alpha decay. Since the proton number decreases by two, any given atom undergoing alpha decay becomes an atom of an element two places below itself in the Periodic Table. In itself, an alpha particle is a very stable particle, and although alpha particles are a type of radiation, they can be absorbed by just a thin sheet of paper or a few centimetres of air.

There are two types of beta particles: electrons and positrons (which are equivalent to an electron but have a positive rather than a negative electric charge). There are also three types of beta decay, in which an unstable nucleus converts into another nucleus with the same nucleon number, but a different number of protons. In two of the three types of beta decay, the proton number decreases by one, converting the original atom into that of an element one place below it in the Periodic Table. In the other type of beta decay, the proton number increases by one so the new element is one place higher on the table of elements. Beta particles need no more than a thin (around 3-millimetre) sheet of aluminium to stop them.

Gamma rays are rather different. They can only be stopped by several centimetres of lead or many metres of concrete. Unlike alpha particles and beta particles, the release of gamma rays does not cause a change from one type of nucleus to another. Instead, the emission just releases excess energy from newly formed nuclei. But it does this super-fast. Gamma rays travel at the speed of light and are on the electromagnetic spectrum along with radio waves, light, ultra-violet and X-rays, albeit at the higher energy end of that spectrum.

As we saw in Chapter 1, the creation of one element from another as a result of radioactive decay was first theorised by none other than Ernest Rutherford and Frederick Soddy in 1902 and is known as transmutation. A nucleus that results from radioactive decay is known as a 'daughter'. This daughter may, like its parent nucleus, be radioactive or it could be stable and will not therefore decay further. If the daughter is radioactive, not only can it decay, but any daughter it might produce can also decay. In fact, this process continues in a so-called 'decay series' until a stable nucleus is formed that is not radioactive.

A life half lived

Clearly, radioactivity is an ongoing process, so how long does a particle or material remain radioactive for? This is indicated by its half-life.

The half-life is the time it takes for half of the starting amount of radioactive atoms to decay. If you measured the radioactivity of whatever number of radioactive atoms you chose to start with and then measured this again at the end of one half-life period, you will find that 50 per cent of them will have decayed. If you measure again after another half-life has passed, half of the radioactive atoms remaining will have decayed, leaving you with a quarter of the initial quota of radioactive atoms. Over the next half-life, an eighth of the original number of atoms will decay, and so it goes on.

One interesting point is that the half-life is statistical, in the sense that it is impossible to tell if one particular nucleus will decay or not within each half-life period. But you do know that overall, a half of the nuclei that you start off with will decay within the period of one half-life.

To calculate a half-life, you measure how much the radiation coming from a radioactive element fades over a relatively short time. It is a good job that half-lives can be determined like this because some are incredibly long. Radium, for instance, has a half-life of 1,602 years, and for plutonium it is 24,400 years, making both of these quite difficult to dispose of.

By contrast, some half-lives can be very short. For example, the half-life of iodine-123 is 13.2 hours, making it ideal for use as a contrast agent in some medical scans.

Technetium-99 is a radioactive atom also used in medicine for imaging the body when searching for cancers or looking at blood flow and it has a six-hour half-life, whereas another medical tracer, Fluorine-18, has a half-life of 1 hour 50 minutes. The latter has the advantage that it does not expose the patient to radioactivity for a long period of time.

There are other radioactive isotopes used to treat cancer, so radioactivity is not always a bad thing.

But the half-lives of these useful medicinal tools are tiny in comparison to the half-lives of materials commonly used as fuel in the nuclear power industry. Some of these materials remain radioactive for so long it is almost impossible to comprehend. Plutonium's half-life of 24,400 years is dwarfed by uranium-235, which clocks in at around 700 million years, while uranium-238 has a half-life of a jaw-dropping 4.5 billion years.

When radioactivity is at a level far above those naturally found all around us and even within our own bodies, it can harm humans and other living organisms by altering the genetic structure of cells, interfering with how cells replicate or even killing cells. Too much exposure can result in radiation sickness or radiation burns or induce cancers including leukaemia. Therefore, measures are needed to prevent radiation from fusion reactors from reaching workers or members of the public. As we saw in the last chapter, these measures include employing specialist shielding materials.

But this does not protect people from the waste products created once components reach the end of their working lives. Clearly, storing the hazardous waste from fission reactors for what in all practical purposes is an eternity is not an ideal situation. Not least because burying waste deep underground – which is the disposal solution of choice for most countries for radioactive material that cannot be recycled – is extremely complex, partly because geology can change over timescales of thousands of years and alters the minute you drill into it. So, if we build fusion power stations in the future, are we not just adding to this unpleasant legacy?

Going to waste

Fortunately, the answer is not to any appreciable extent. As mentioned in the introduction, one of several advantages of a fusion power plant over a fission power plant is that fusion reactors do not produce long-lived radioactive waste.

For starters, all that is left over from the used-up fuel in a fusion reactor is helium, which is not radioactive. This helium by-product diffuses out of the plasma and is drawn out of tokamak-design fusion reactors via the divertor component (which we learned about in Chapter 3). The divertor therefore acts a bit like a giant ashtray, collecting the waste helium. Over time, the tungsten surface of the divertor – which in the case of ITER is made up from a series of cassettes arranged in a circle – degrades due to erosion by the plasma. But during a shutdown period for the reactor, degraded cassettes can be replaced.

Although the helium itself is not radioactive, the divertor cassettes will be too radioactive for humans to handle thanks to their interaction with the neutrons released in the fusion reaction. Therefore, the cassette-replacing procedure would be carried out using robotic handling systems within an 'active cells' facility (also known as a 'hot cells' facility), which we will look at in the next chapter. A hot cell has doors at either end with tracks running through it. This arrangement allows a door into the reactor to be opened, a component to be pulled into the hot cell along the tracks and that door to be closed before the door at the other end is opened to release the component.

Like the divertor, the steel walls of a tokamak also become radioactive as a result of being constantly hit by highly energetic neutrons. This means the reactor vessel is equally off-limits to human workers during shutdown. But this radioactivity is short-lived. The half-life of tritium, used to fuel the fusion reactor, is 12.5 years, and the half-life of the other radioisotopes produced in the ITER fusion reactor are similarly short-lived. What this means is that fusion waste could potentially be left in place after a reactor has reached the end of its lifespan. The half-lives of its irradiated components are such that the radiation would be decayed to safe levels within 100 years. So, a century after they had been created, these waste products would have a sufficiently low level of radioactivity that they could be recycled for use in other fusion reactors. Humans could even walk into the reactor after 100 years had passed as there would be no remaining radiation hazard. Fusion does not, therefore, have the long-lived radioactive waste legacy that nuclear fission plants create.

Most of the problems from the current crop of nuclear reactors come from the spent fuel, often uranium-235, which is highly radioactive and extremely hot. The fuel rods end up containing a complex radioactive inventory, including krypton, xenon, strontium and yttrium, which are hard to separate chemically. Since it can take hundreds of thousands of years for high-level radioactive waste such as spent fuel rods to decay to safe enough levels to handle, in the meantime, it must be stored safely either in special pools or inside a dry cask. The pools enable the spent fuel rods to be immersed at least 6 metres (20 feet) underwater. This is enough to protect any nearby workers from the radiation. Casks are generally made from steel and house spent fission fuel that has already been cooled for at least a year in a pool.

Each cask contains an inert gas to surround the fuel, while its outside is clad with more steel, concrete or another material which is able to shield anyone from the radiation contained inside. The underground vaults these casks are ultimately placed into for long-term storage are also made from materials able to shield workers and the general public, such as concrete and metal.

Having a smashing time

Of course, it is not just storage that is a problem when it comes to nuclear waste. It must be transported around, and this has its own challenges. One of the most dramatic demonstrations of just how stringent the safety measures for transporting nuclear waste are took place on 17 July 1984 in the UK. A Class 46 diesel locomotive pulling three coaches rounded a bend at 100 mph (161 kilometres per hour) and ploughed head-on into a nuclear flask truck. As the smoke from the explosive impact cleared, it revealed catastrophic damage to the locomotive, and the coaches behind it had derailed. But this was no accident. It had been a deliberate crash. Not by saboteurs or terrorists but by the then Central Electricity Generating Board (CEGB).

With 'Operation Smash Hit', the CEGB were trying to prove the point that the nuclear flasks they regularly used to transport spent fuel rods from nuclear power stations all around the UK to the Sellafield reprocessing plant in Cumbria were a safe method of transportation.

The train, which had been driven under automatic control, and had gradually accelerated up to 100 mph from its starting point in a siding just over 8 miles away, was designed to replicate a common passenger train of that era. The weight of locomotive plus carriages was 244 tonnes (269 US tons), and they were destined to smash into a 51-tonne (56 US ton) flask.

This flask was then the standard container for spent fuel rods, which once they had been cooled down for 90 days after coming out of a reactor were loaded underwater into steel skips in batches of 200 at a time. Each skip was then placed inside a flask with 14-inch-thick steel walls. Sixteen bolts were used to seal the lid of the flask down, each bolt capable of withstanding a load of 152 tonnes (168 US tons), before the flask was fitted on a type of flat-bed truck.

For the purposes of the test, 1 tonne of water plus 200 steel bars were put inside the flask to replicate the normal load of 200 used fuel rods. To simulate a real-life worst-case scenario, the truck carrying the flask was deliberately de-railed and tipped up onto its side.

After a few minutes of delay while the Police dealt with nuclear protestors, the test got under way. Soon, the 1,500 invited guests, including emergency service chiefs, MPs, local councillors and press, could see the unmanned diesel steaming around a bend in the track coming into view. The impact, as it smashed into the flask, was recorded from a range of different angles by 32 cameras (see link to the video in the Further Reading section), including one on board a helicopter filming the train's entire journey from above and a slow-motion camera sited to the side of the impact zone. By anyone's standards, the crash was dramatic. But did the flask survive intact?

The CEGB engineers had to wait a while after the impact until they had the all clear from the emergency services before entering the crash zone. Only then could they carry out the detailed measurements designed to put to bed any safety fears surrounding transportation of nuclear material.

It was soon clear that, despite some surface scarring of the steel walls and buckling of the cooling fins on the outside, the flask had indeed remained intact. Before the test began, the flask had also been pressurised to 100 lbs per square inch and, when the engineers measured the pressure inside the flask after the impact, it had only decreased by 0.26 lb, which would pose no risk in real life.

But although transporting nuclear materials can be done safely, as this spectacular demonstration showed, it is complex and expensive to do. Therefore, it makes sense to either avoid or minimise the amount of moving around of radioactive materials.

Captive breeding programme

Fortunately for fusion energy, there is little need for transportation of nuclear fuels because inherent to the design of several of the tokamaks under development is what amounts to a 'fuel factory' in their reactor walls.

These fusion reactors use deuterium and tritium as fuel. Deuterium is extremely plentiful as it is found in water. This means you can take it out of seawater and thereby have an abundance of that fuel for hundreds of millions of years. On the other hand, tritium is so scarce in nature that future fusion plants would not be able to source it in large quantities. The solution scientists and engineers have come up with is to make tritium within the reactor.

To achieve this at ITER, lithium will be put into 'test blankets' lining the beryllium walls of the fusion vacuum vessel. Some of the neutrons that hit the wall will then create tritium out of lithium in a transmutation process. When the tritium gets knocked out of the nucleus of the lithium it is in the form of a gas. So, you get bubbles of tritium that gather together in the lithium within the vessel wall. This can be sucked out, then put back into the machine to fuel the fusion reaction. This process is known as 'breeding' tritium, and ITER will test out different designs for breeding blanket systems to determine which is the most successful.

Because tritium is a gas, it can get into tiny crevices, creating tritium-active components that retain radioactivity over time. In addition, as the lithium gets used up during the breeding process, this means the radioactive wall components containing lithium will have to be regularly taken out and replaced. This necessitates a specialist handling system, including a hot cell complex within a fusion reactor's tritium facility, as well as sufficient shielding to safeguard workers from the radiation. Although this is a reasonably complicated engineering solution, on the plus side, not only are you breeding your own fuel to help with supply issues, there is also no question of the fuel from a fusion plant becoming waste.

Another important research area within the fusion field is the development of so-called low activation materials, which do not become highly radioactive in the first place. This is the ideal scenario, since it reduces the need for costly and technically demanding handling or storage solutions.

Recycling scheme

For existing nuclear power plants, which rely on nuclear fission to work, fuel is also a precious commodity, and recycling spent fuel has several benefits.

Firstly, it simply makes sense to get the most out of the fuel as possible, and thanks to reprocessing techniques, unused plutonium can be recovered from spent uranium-238. This is achieved by dissolving the fuel rods in concentrated nitric acid, then using solvents to extract both plutonium and also uranium from the resulting solution. This reprocessing enables 25–30 per cent more energy to be gained from a given amount of uranium fuel than could be produced with just a one-time use.

Secondly, as disposing of high-level radioactive waste is such a headache, getting rid of about four-fifths of that waste by reprocessing is extremely helpful. Another advantage of reprocessing is that the waste which this processing leaves behind has a much shorter half-life than unprocessed spent fuel.

It is therefore small wonder that reprocessing is carried out by several countries around the globe, although it is by no means a universally adopted practice. In the UK, at the Sellafield nuclear site in Cumbria, spent nuclear fuel is reprocessed by extracting plutonium and uranium. Sellafield also handles low, intermediate and high-level waste prior to storage. In the United States, the H Canyon chemical separations facility in South Carolina handles spent fuel rods, recovering uranium from them. Meanwhile, the Orano la Hague site in France has been handling and recycling French nuclear waste since 1966.

By contrast, the first commercial reprocessing plant in Japan, the Rokkasho Reprocessing Plant, is not yet fully constructed. Once operational, the site will be able to handle up to 800 tons of uranium a year, which is approximately the amount of spent fuel from 40 reactors. In the meantime, Japanese nuclear reprocessing takes place at sites in Europe.

Which storage and safe disposal facilities for the radio-active waste from future fusion plants will be used, or indeed if further storage and recycling sites will be built, is a matter for host nations to decide in due course. But wherever the waste is sent, it is certainly the case that there would be a lot less of it than currently comes from fission reactors, and it would be less complex to deal with.

As we have discovered, though, what waste there is still requires very careful handing. To understand more about how to safely extract the radioactive material from the insides of fusion reactors, we now need to turn our attention to the world of robotics.

TO INFINITY AND BEYOND

7

Rise of the robot dogs

With its ability to walk over challenging surfaces, nimbly climb stairs and even dance thanks to such a high level of motor control, many of us have fallen for the charms of Spot®, the dog-like quadrupedal robot created by US engineering and robotics company Boston Dynamics. More importantly for a nuclear environment, Spot can be fitted with a range of different sensors and go into areas too hazardous for humans.

In September 2021, two Spot units were tested at Sellafield at the Calder Hall nuclear power station, which was in the process of being decommissioned. Calder Hall's construction had begun in 1953 and it operated for 47 years, before being closed down in 2003.

Decommissioning nuclear power stations is a lengthy and challenging process, mainly because of the levels of radioactivity in various areas of the site. Spot is able to get itself around a variety of different terrains without falling over and can be equipped with mapping capabilities and radiation detection sensors. This enables the robot canine to be sent on inspection missions into areas that would be unsafe or very complex for humans to enter, and to either work autonomously or be controlled remotely by engineers. If Spot was used to carry out routine inspections, it could completely remove the need for humans to carry out such tasks, thereby reducing any risk to personnel and decreasing the time they need to spend on checks. As Spot could also perform more frequent checks and so flag up required maintenance in a more timely fashion, using such robots should save money.

Although at the time of writing this testing is taking place in a fission facility, the hope is that Spot would also prove useful in a fusion environment, carrying out inspection roles. Spot's employment within the JET facility is also being considered, since it could be used to retrieve material samples from around the reactor vessel at times when radiation is present around the reactor.

Remote working

Less entertaining to watch, but vital for fusion reactors, are the sorts of advanced robotics that can maintain the plant and fix components in areas people cannot enter. This, in other words, is a remote handling system, which is a form of robotics. The key distinction between the types of robotics being whether there is a person in the loop.

If we think about the robot arms used in car assembly lines, for example, there is not a person in control. This is because these robots are either pre-programmed or have enough sensors that, thanks to the teaching they have previously received, they can change their behaviour to suit new situations. A person-in-the-loop system is more akin to robotic surgery. In this case, the surgeon is operating a piece of equipment that has surgical instruments at the patient end of it. This type of robotics is very useful in surgery as it can remove any natural tremor from the surgeon's hands and also move any instruments inside the patient's body much more precisely. Robotic surgery systems effectively have a scaling factor between the surgeon's movements outside the patient's body and what is happening on their insides – larger movements made by the surgeon translate into much finer movements at the other end. Hence the system acts like a form of mechanical microscope.

With person-in-the-loop robotics in a nuclear environment, the main benefit is being able to protect workers from hazards as they are no longer required to go into areas where they would receive a radiation dose. But there is also another benefit in that you can design-in capabilities for your tool-and-person system that neither would have on their own. For example, a person can pick up 100 kilograms (220 lbs), but with the right design of remote handling system, that weight could feel as though it is 5 kilograms (11 lbs). So, the robotics can enhance the capabilities of the person by magnifying their natural abilities.

Hostile environment

Robotics is necessary in a fusion reactor not only because there are chemical hazards from materials, such as beryllium, but also because fusion creates so much energy. As we saw earlier, for deuterium and tritium fusion, the energy created is primarily in the form of high-energy neutrons and gamma radiation. And although the high-energy neutrons only last while the fusion reaction is taking place, they leave behind a radioactive legacy. (The neutrons themselves do not pose any hazard for people thanks to the shielding described in the last chapter.) In fact, the radiation levels coming from the materials that the neutrons activate are so high that once a reactor has been up-and-running it is impossible for people to ever go inside the reactor vessel. Humans can only enter while construction of the reactor is taking place, presenting a problem for the ongoing maintenance and safety checking of fusion reactors.

The former will be essential as some components of the machine will simply wear out over time, or their replaceable nature is an integral part of the reactor design. The latter will be a result of future regulation of fusion power plants, which will demand stringent safety checks. But if people can't carry out the required maintenance and safety tasks, what types of robotic solutions could be used instead?

Over the past couple of decades, UKAEA engineers at JET have been developing a robotic remote handling system tailored to their fusion environment. This does not use fully automated robots like Spot. Instead, the JET system works in a similar way to robotic surgery.

When irradiated components are removed from the inside of fusion reactors, they can irradiate further machine components on their way out. This means the radiation effects are not just confined to materials within JET's main reactor vessel. Fortunately, these irradiated materials are not necessarily nuclear waste, in the sense they may not actually be disposed of. But they will at least require reprocessing or recycling.

So, some of the robotics used in a fusion environment such as JET can be designed to work inside the vessel itself, while further robotic systems can be employed outside. According to Rob Buckingham, head of the UKAEA's robotics division, Remote Applications in Challenging Environments (RACE), when fusion reactors are initially designed, the following question needs to be posed: 'Is there any point where you would decide to have people in protective clothing walking around, or do you have a suite of robots to operate and maintain the machine?'

He explains to me that there is a big difference between designing for remote maintenance and designing for human maintenance, simply because robots do not have to be the same strength or shape as people. Bespoke robots can be designed for specific tasks and spaces, so they are essentially a remotely operated tool for completing a task – albeit a very smart tool.

The remote handling system was put to the test in the early 2010s when the inner carbon wall of JET was stripped out and replaced with a new lining of beryllium and tungsten. This was to mimic the wall that ITER will install and carry out experiments to determine what erosion and other effects, that have an impact on the lifetimes of components, are likely to occur once ITER is operational. Ninety per cent of the wall replacement work was carried out using the RACE remote handling robotics, with the remainder done directly by people.

As part of the system development, the RACE engineers adapted ideas from space research and other industries working in environments where radiation and contamination control is a challenge. At JET they needed to ensure that when the reactor vessel was opened up and entered, contamination was not spread around. This contamination comes mainly from the tritium fuel that goes into materials. In fact, how to safely handle tritium is such a challenge that it is the basis for one of the largest research programmes at JET.

Three's a crowd

To address the issues surrounding the safe handling of tritium, the RACE engineers had to first think about radiation in two ways: what does it directly destroy and where does it go and cause problems? Then they had to take into consideration the specific architecture of the reactor and the spaces they needed to reach. These factors impose constraints on the types of robotic machines you can use. While a simple crane, for example, which can handle heavy components precisely and pull them out through a door into the reactor, would be relatively easy to create, life inside a tokamak fusion reactor vessel is not so simple. With an intricate doughnut-shaped machine full of oddly shaped nooks and crannies packed with expensive equipment that must not be touched, you need a more complex robotic solution.

This meant developing robots with relatively slender arms that can reach into awkward spaces to pull out or manipulate large, heavy components. While in an ideal world you would make robots thicker and stiffer so that they are as strong as possible, in this case, this was impossible because there was simply not sufficient space available. Despite this restriction, the RACE robot is more than up to the lifting and manipulation tasks required, in part thanks to a robotic gadget called MASCOT, which is connected to its reactor end. (MASCOT was originally developed in Italy, and the acronym comes from the Italian: *MAnipulatore Servo COntrollato Transistorizzato*, which translates as 'Servo Controlled Manipulator with Transistors'.)

The robot is operated by an engineer sitting in a control room moving two arms with grippers. The overall system works in a similar way to the funfair game in which you try to use a crane-like manipulator arm to grab a toy. In the case of the JET system, the manipulators have force feedback which enables the operator to feel exactly what is happening on the other end of the arms connected to the MASCOT device. MASCOT can be used to fetch different tools from a selection on a trolley inside the reactor, then directed to the part of the machine that needs to be worked on. Since all these processes are being initiated from a control room about 20 metres (66 feet) away via the movements of a person, the system is very similar to robotic surgery.

Care provider

While these types of maintenance robots work to protect our health, we also need to protect their health. A robot cannot just be placed into a high-radiation environment with no issues. The robot will be affected by the radiation, so creating resilient robots that can work in harsh environments is very important.

Research into developing such robots is currently being carried out by the space industry as well as in the nuclear industry. In space, not only is there radiation but any robots deployed in planetary orbits also need to be able to cope with impacts from cosmic rays (which consist of high-energy atomic and subatomic particles). Cosmic rays can play havoc with the electronics if they flip the state of a bit of memory, for example, and this could cause the robot to malfunction.

But while cosmic rays are not a concern in a fusion reactor environment, high-energy radiation damages all materials, whether in space or here on Earth. To help combat this, new radiation-tolerant materials are constantly being investigated by the RACE team. However, since new radiation-tolerant materials can often be prohibitively expensive and don't tend to come along too often, the RACE team have been focusing much of their recent efforts on condition monitoring and control theory – the mathematics that govern robot control systems. This must take into account not only how the robot works, but also any potential disturbances to its behaviour caused by the environment it is working in, including degradation in performance due to age or radiation. The more accurate the mathematical model, the better you can control your robot. The RACE team are aiming to understand more about control theory so they can get the optimum functionality from the robots they use and find the best ways to control large flexible robotics.

Knowledge transfer

While these projects are looking to the future, in the here and now the UKAEA are seeking to spin out their technologies to industries with similarly challenging environments, such as other types of nuclear facilities, petrochemical plants, mining and the construction industry. The robotic systems developed by its RACE division, for example, are applicable to any research or industrial environment where it is better to send in robots than people.

One facility already benefiting from this know-how is the European Spallation Source (ESS) based in Lund, Sweden, with whom the UKAEA are working to develop remote handling capabilities. The ESS is a multinational facility for neutron scattering experiments scheduled to open for users in 2023. Once the facility is operational, the highest ever intensity proton beam will be fired into a spallation target made of tungsten. Spallation is a nuclear reaction in which nuclei are hit so hard with other nuclei or high-energy particles that they disintegrate into their constituent protons and neutrons. It can occur naturally (in space) or be created in a laboratory. At ESS, the protons in the beam will have such a high energy that they will blow the tungsten atoms of the spallation target apart, causing neutrons to be emitted in all directions. These neutrons will then be used to study complex physical, chemical or biological specimens.

Over time, the target will become brittle, and after a few months it would start to crumble apart. To combat this, the ESS target design is a wheel with 36 sectors that look a bit like a flatter version of the blades on a child's hand-held windmill toy. This wheel will rotate in the proton beam so that the radiation damage is spread out and the target therefore lasts for several years rather than several months. Even so, it will need replacing. Plus, there will be fifteen separate beamlines to maintain, and not only high-energy neutrons but also gamma rays and X-rays are produced by the spallation process. Despite steel, lead and concrete shielding providing protection from the radioactive isotopes, and despite boron carbide and cadmium being used to absorb low-energy neutrons that would otherwise activate surrounding materials, there will be areas of the ESS facility with levels of radiation dangerous to humans. Therefore, workers will not be able to simply go in and replace the targets. This is where robotic handling comes in.

The robotic handling will work within an 'active cells facility'. This is basically a room that takes the highly irradiated targets from the ESS machine, cuts them up and puts them in boxes ready for safe disposal. The ESS will generate approximately the same amount of radiation dose that a fusion reactor would create, so its materials will have a similar level of radioactivity. This means that when designing the system, it was important to think very carefully about what electronics could be used and how the robotics would need to be shielded within the active cells room.

Development of such remote handling robotic systems is acting as a test bed for future fusion applications, as well as generating additional value from the UK's fusion research programme. At the UKAEA, unsurprisingly, the main aim for their fusion research is to get fusion energy delivering power into the national grid. But looking out for any knock-on benefits for other areas of scientific research and for other sectors in the economy is also an important part of their mission, as we will learn more about in the next chapter.

This type of knowledge transfer is by no means an unusual occurrence. Spin-offs from the space industry, for example, have found their way into our everyday lives. NASA-funded research alone has helped to give us a whole range of familiar items, including cordless drills and cordless vacuum cleaners, the metallic foil blankets used by hikers, runners and emergency services and even the memory foam inside the pillows that many of us sink our heads into each night.

Only time will tell whether fusion research will provide a similar array of everyday products. But even if we are not going to be tracking down fusion-inspired goods in our favourite stores anytime soon, what else could fusion research lead to?

Out of this world

While the US Navy have now abandoned their 2018 patent for a 'plasma compression fusion device' that could form the engines for their future warships, fusion drives are looking like a potentially promising prospect for space travel and exploration.

Space exploration would benefit from fusion power because the fusion rocket engines under development could provide a path towards faster journey times, opening up more possibilities. With present-day technology, manned interplanetary travel would at its best be extremely time-consuming. A trip to Mars, for example, would take seven to nine months depending on the position of the planet in its orbit. Mars is around 62.07 million kilometres (38.6 million miles) away from Earth when it is at the closest point in its orbit to us. Getting humans to planets further away than this, such as Saturn, which is 1.2 billion kilometres (746 million miles) away when closest to Earth, is just not feasible with existing spacecraft propulsion systems. Unless, that is, you would be happy being cooped up in a tiny spacecraft for eight years in order to get there. But what if faster spacecraft engines could be developed that enabled journey times to be slashed?

This was the question posed by NASA's Fusion Driven

Rocket (FDR) project, which ran in 2011 and 2012 and progressed thanks to two early-stage research studies developed through the NASA Innovative Advanced Concepts programme. The rocket concept built on work carried out by research company MSNW LLC, based in Redmond, Washington, on creating the conditions for fusion via the magnetically driven implosion of metal foils on a magnetised plasma target. In NASA's extension of the MSNW work, the metal foils and metal shells acted as a propellant as well as producing the conditions for fusion.

The propulsion system of NASA's FDR rocket design contained several metal liners that would be movable thanks to electromagnetic induction. Once triggered to move, they would converge together, forming a thick blanket around the plasma, compressing it enough for fusion to occur. The energy released when the fusion reaction is operating is in the form of radiation, neutrons and other particles. Almost all of that energy would be absorbed by the metal blanket formed from the metal liners.

The amount of energy absorbed, in addition to the heat created when the compression reaches its peak, would be enough to vaporise the metal blanket, turning it into an ionised gas. It is this ionised gas that would become the propellant for the rocket.

In general, to create thrust from a rocket engine, gas is heated or put under pressure and then forced out through a small nozzle. Thanks to Newton's laws of motion, the action of the gas thrusting out of the nozzle creates an equal and opposite reaction, pushing the rocket in the opposite direction.

In the FDR rocket engine design, the ionised gas from the vaporised metal blanket is expanded through the nozzle. This results in higher levels of thrust for propelling the rocket forward than from existing rocket engines. The absorption of the fusion energy by the metal blanket also has the handy side effect of protecting the rocket from the extreme conditions of the fusion process.

One of the most useful features of this FDR system is that, unlike the fusion reactors experimenting with terrestrial electricity generation, these fusion devices will not generate electricity. Instead, the energy released from the fusion reaction goes straight into making the spacecraft's propellant.

In the NASA design, the fuel is solid lithium, which is relatively compact to store. The fusion energy quickly heats up the propellant and accelerates it to an exhaust velocity of more than 30 kilometres per second (67,000 mph). While the exhaust velocity is not the only factor that determines the forward speed of a rocket, it would nevertheless provide the ability for rocket engines fast enough for further manned exploration of our solar system and potentially interstellar travel.

Having drive

It is this promise of faster journey times to planets in our solar system, and the eventual move of humans into space, that is driving Helicity Space. This private company, founded in 2018 and based in California, is working solely on fusion for space propulsion.

Also working on fusion systems for space propulsion is Bletchley, UK-based Pulsar Fusion. The company was founded by entrepreneur Richard Dinan, known to many British TV viewers from his appearance in the reality series

Made in Chelsea. Dinan, who entered the fusion sector in 2013 due to a 'passion for the physics', soon built up a team of experienced scientists and engineers to work on fusion projects and related technologies.

Pulsar Fusion's rocket programme includes seeking to develop a Direct Fusion Drive (DFD), which could provide the thrust and electric power for long-distance spacecraft travel. At the time of writing, the company's team of thermonuclear physicists are still in the design phase of this project, but they have already carried out successful small-scale tests of their modernised take on a 1970s' NASA design of a satellite thruster. This uses plasma heated to several million degrees Celsius to produce thrust. Pulsar Fusion's first trials of that technology, in October 2020, involved a prototype engine that runs on a xenon and krypton fuel, and they managed to achieve exhaust velocities of 27 kilometres per second (60,000 mph). While this is a competitive and efficient exhaust velocity for an engine of that size, their DFD project aims to achieve an exhaust velocity a thousand times faster. This would potentially allow for very rapid acceleration through space since it is the exhaust velocity in conjunction with the decreasing mass of a rocket as it uses up fuel – and in some cases jettison stages – that determines the thrust of a rocket. While the thrust is greater than the sum of the rocket weight and the drag exerted on it, a rocket will continue to accelerate.

At the time of writing, Pulsar Fusion recently gained experience of moving from the concept stage to real-life rocket trials when in November 2021 a separate team of combustion rocket scientists tested a hybrid rocket design. This rocket burns a mixture of nitrous oxide and recycled high-density polyethylene (HDPE) and was tested at the

COTEC Ministry of Defence base in Salisbury, which offers a range of specialist testing facilities for explosives- and weapons-related research. (To test fire rocket engines, COTEC has an area where they can be mounted sideways and firmly secured in a static arrangement.) Due to the success of these Salisbury tests and other testing in Switzerland, Pulsar Fusion intends to carry out trials using liquid propellant in 2022, while work continues simultaneously towards the construction of their DFD fusion rocket.

While, as we are about to see in Chapter 8, billions of dollars are now pouring into both public and private fusion projects, this was not always the case. In fact, the greatest challenge in the early days of Pulsar Fusion, says Dinan, was raising capital. 'Ten years ago it was almost impossible to raise money in fusion. Now, thankfully times have changed,' he explained to me, adding that the most significant milestone for him so far has been hearing the HDPE rocket firing in the mountains in Switzerland. This caused him to pause and reflect: 'So many things had to go our way in order to make that possible. Many companies talk about designing rockets but few in this part of the world have actually built and tested two, in two countries, in a snowstorm!' For that moment, Dinan says he felt 'like anything is possible'.

THE RACE IS ON

8

Urgent assistance required

We have seen over the past few chapters just what an extraordinary effort has been put in by the scientists and engineers striving for a fusion-powered future via a range of different technological approaches. Back in Chapter 2, we also learned about the long history of fusion research stretching back to the 1940s. Since we still do not have any electricity from fusion on the grid, and much more work is required to reach that point, it seems fair to ask: is this all really worth it? To help answer that question it is useful to first turn to some statistics on the power we use.

With the exception of a few dips – including after the 2009 financial crisis and during the early stages of the Covid-19 pandemic – global energy consumption has been rising year-on-year for over fifty years. Looking solely at electricity use, in 2020 the global demand fell quite considerably – by up to 30 per cent in some places – during the pandemic lockdowns when business, transport and industry were restricted.

But as these activities resumed and economies began to recover, the upward trend for electricity usage resumed. This was bolstered not only by pandemic recovery but also by the fast growth of some emerging economies.

Hearteningly, 2020 also saw a 3 per cent increase in the global use of electricity from renewable sources. But with a quarter of the world's greenhouse gas emissions coming from non-renewables in the power sector and energy consumption continuing to rise, developing more green electricity-generating options could not be more important. In fact, if the world wants to reach the goals from the Paris Agreement, new, clean power solutions are essential.

The Paris Agreement was adopted by 196 parties (nation states and other regions) at COP21 (the 21st UN Conference of the Parties) in 2015. It is a legally binding international treaty, with a goal of limiting global warming to 'well below 2, preferably to 1.5 degrees Celsius, compared to pre-industrial levels'. It entered into force on 4 November 2016. But progress towards its aims has been slow. To meet the Paris target, the move away from coal towards clean power needs to proceed five times faster than the rate at which this was happening in 2021.

It was against this background that delegates met in Glasgow for COP26 in 2021. One of the outcomes of COP26 was almost 200 nations agreeing to the Glasgow Climate Pact, which included limiting global warming to 1.5°C (2.7°F) and finalising some outstanding aspects of the Paris Agreement. As one of the many points addressed, the Glasgow Climate Pact 'expresses alarm and utmost concern that human activities have caused around 1.1°C [1.98°F] of warming to date, that impacts are already being felt in every region, and that carbon budgets consistent with achieving the Paris Agreement temperature goal are now small and being rapidly depleted'.

It was realised at the time of the Paris Agreement, back in 2015, that to achieve the target for limiting global temperature rise, countries would need to become climate neutral by the middle of the 21st century. As discussed at COP26, generating power from coal is the single largest contributor to global warming. This led many countries at the conference to sign the 'Global Coal to Clean Power Transition Statement'.

This statement recognises the need to rapidly move away from coal power to cleaner forms of energy, as well as reducing carbon dioxide emissions from any coal that is burned. Of course, to do the former, we will need to find other ways of generating electricity. It is no surprise that one of the commitments contained in the statement is 'to rapidly scale up technologies and policies in this decade to achieve a transition away from unabated coal power generation in the 2030s (or as soon as possible thereafter) for major economies and in the 2040s (or as soon as possible thereafter) globally'.

Advocates for fusion energy hope that it is this technology that will play a major role in the drive to reduce the world's carbon emissions and provide a sustainable power source for centuries to come. But as we have seen, the fusion journey has already been a long one. It is also expensive.

Costing the Earth

While the need for new forms of green energy could not be more pressing, there is no getting away from the fact that the development of fusion energy has a substantial financial cost.

In a survey of fusion energy businesses globally, published in October 2021 by the Fusion Industry Association and the UKAEA, the 23 companies that responded to the survey had received a combined total of nearly $1.9 billion of private funding from their inception to the second quarter of 2021. This was in addition to the $85 million in total they had received in funding from governments.

Not that this was entirely for terrestrial power generation. For the 23 companies included in the survey, while almost all (95.7 per cent) cited electricity generation as their target market, nearly half (47.8 per cent) were aiming for developing viable space propulsion systems and just over a quarter (26.1 per cent) were targeting the marine propulsion sector, aiming to develop fusion-based engines for ships. A combination of medical applications, off-grid energy, clean fuels and industrial heating also made up the stated target markets for their research.

Breaking down these figures a bit, according to the BloombergNEF research group, during 2020, US investors committed $300 million to around 24 private fusion companies based in North America or Europe. In addition, the US Department of Energy (DOE) have been funding US fusion research to the tune of roughly $600 to $800 million annually for decades. Although in recent years the funding had been declining, in 2020 there was additional spending given in the US spending bill, resulting in $671 million going in the direction of fusion energy research.

Also in 2020, the Fusion Energy Sciences Advisory Committee in the US published 'a long-range plan to deliver fusion energy and to advance plasma science'. This report was a fusion community-led strategic plan that identified what was needed in order to realise a fusion energy future.

It highlighted critical areas for future research and development as well as where to prioritise funding in order to advance fusion energy over the next ten years. While the aims include striving for new scientific discoveries, as well as a programme to translate findings from fusion research into industrial applications, the main goal is to deliver fusion energy to the grid.

The US strategy includes building an American pilot fusion power plant by 2035–40 – a project which was the subject of a 2021 National Academies of Sciences, Engineering and Medicine feasibility and planning study – as well as continuing to support partnerships. These include the multinational ITER project in Cadarache in France, as well as private sector fusion companies, which in the last ten years have received $2 billion of US government financial help. This is partly distributed within the DOE's Innovation Network for Fusion Energy (INFUSE) programme, which aims to 'accelerate basic research to develop cost-effective, innovative fusion energy technologies in the private sector'. INFUSE not only offers funding but also access to technical expertise and capabilities at DOE national labs and US universities. Meanwhile, private companies developing technologies not yet mature enough to attract private sector investment can apply for Advanced Research Projects Agency–Energy grants to get similar technical expertise access in addition to financial support.

The private sector offers the prospect of specialist capabilities, and a faster route to commercialisation of fusion power and any related technologies. It also shoulders some of the costs and, due to its inherently entrepreneurial approach, works to tight timescales. This means similar partnerships to the one between NASA and SpaceX (the private space travel company founded by Elon Musk in 2002), which combine the advantages the private sector can bring with decades of scientific experience from national institutions and other public sector facilities, are likely to be forthcoming within the US fusion sector. This will mark a sea change: going from the government solely funding programmes that advance scientific knowledge about plasmas and fusion but have constraints on duration and how the money is spent to also funding the commercialisation of fusion. In this final phase, much of the focus will have to be on becoming a viable player in the energy marketplace. This will require a different skill set and mindset compared with purely academic research, which the private sector is able to provide.

The fusion strategy also seeks to redirect US DOE Fusion Energy Sciences research programmes in other directions. This is in order to facilitate delivering fusion energy as swiftly as possible and recruit more talent into the fusion sector.

To reach the goal of building a pilot power plant, the strategy outlines the three major scientific and technological areas that need addressing. The first is developing the science and technology needed to both confine and also sustain a burning plasma. Second, materials need to be developed that can withstand the extreme conditions faced in a fusion reactor. Third, the fusion power created needs to be harnessed. The latter will involve developing the ability to breed fusion fuel within the pilot plant's reactor, as well as engineering the technologies needed to generate electricity from the fusion.

The report recognises that in order to achieve these advances, in addition to progressing fundamental plasma science and the data processing required to support fusion, it will be critical for the United States public fusion sector to work with and draw on the work of its international partners as well as its partnerships with private fusion companies. It is certainly a timely strategy given that, as at 2020, 35 per cent of US energy consumption came from petroleum, 34 per cent from natural gas, 10 per cent from coal, 9 per cent from nuclear fission plants, and only 12 per cent from renewable forms of energy.

Together in electricity dreams

The United States is not the only nation to put a roadmap for fusion in place. In October 2021, the UK government published 'Towards Fusion Energy', a paper outlining its fusion strategy. This looks at not only how fusion could help meet the UK's Net Zero target by 2050, but also at energy demand well beyond that date.

According to the document, the two overarching goals are 'for the UK to demonstrate the commercial viability of fusion by building a prototype fusion power plant in the UK that puts energy on the grid' and 'for the UK to build a world-leading fusion industry which can export fusion technology around the world in subsequent decades'. These goals are to be met by working with the UKAEA, which manage the national research programme on fusion and as a consequence receives the majority of the UK government's funding for fusion energy development.

While it is by no means its only function, fusion is a major focus for the UKAEA. This is evidenced by the fact that the largest slice of the approximately £110 million ($138 million) overall government funding that UKAEA received in 2020–21 went into experiments either directly on fusion or related to it. Since it is critical for the fusion field to create sustainable skills in the current and future industry workforce, the annual government funding for the UKAEA also covers training and apprenticeships, as well as the running of the research facilities for UK and international users.

When it comes to spending public money, unlike financing in the early days of fusion (which we learned about in Chapter 2), nowadays there is never a money-is-no-object approach to fusion experiments. At present, the UK government funding not only has a set budget, but it is also being channelled into applied projects that have constraints. For instance, the funding for the Spherical Tokamak for Energy Production (STEP) programme has the specific goal of having a prototype fusion power plant operating in the UK by 2040 and supplying power into the national electricity grid. All the research within the STEP programme, therefore, is targeted at unblocking the challenges in the path to creating a viable power plant using tokamak technology. Similarly, the UKAEA also have a project under way that looks to spin-out key fusion technologies to industry. This includes the robotics discussed in Chapter 5, which can help other industries that work in challenging environments.

Most of the fundamental research and blue-sky thinking on fusion that takes place at the UKAEA, such as looking into methodologies and ideas which are not part of the separate government programmes, are funded by the Engineering and Physical Sciences Research Council (EPSRC). In contrast to the government-funded STEP programme, part of which aims to develop intellectual property and investable concepts, prototypes and patents for the UK, thereby requiring much of the research progress to be kept under wraps, the

EPSRC-funded research is shared internationally and published in scientific journals.

The UKAEA's international collaborations are also research council-funded. These include the UK's participation in EUROfusion.

EUROfusion is the European Consortium for the Development of Fusion Energy, which was formed in October 2014 by fusion research bodies from European Union member states, as well as Switzerland, and at the time of writing it still includes the United Kingdom. Its purpose is to strengthen European collaboration on fusion research by supporting and funding research on behalf of the European Commission's Euratom programme.

Such collaboration is seen as vital to the future success of bringing fusion technology to maturity. 'I think everybody recognises that the challenge is just too big for anyone to do on their own,' stated UKAEA head of the executive office Nick Walkden in a video call with me.

There are also large sums of public money being put towards multinational and national fusion research by other countries around the world. In 2012, for example, the South Korean government announced it would spend around $941 million on its Korean Demonstration Fusion Power Plant (K-DEMO) project.

The Korea Institute of Fusion Energy, which developed and operate the tokamak research reactor KSTAR, are not only looking to create core technologies that will enable the commercialisation of this green form of energy. They are also looking to secure their own energy supply nationally, rather than relying so heavily on importing energy from overseas.

China is also investing heavily in nuclear fusion, with almost $900 million of funding being spent in 2019 on the building and operation of the Experimental Advanced Superconducting Tokamak (EAST). The Chinese government has since matched that funding in another award to the EAST project.

Meanwhile, the financial cost of getting ITER to the first plasma stage is approximately €17 billion ($18.2 billion). In addition, the costs for ITER's deactivation (which is scheduled to take place between 2037 and 2042) and decommissioning phases were in 2001 calculated to be €281 million ($301 million) and €530 million ($568 million) respectively. This is on top of the costs that will be incurred during the operational lifetime of the machine.

Collaborate or go home

As we have seen, internationally, the fusion race is a mixture of collaboration and cross-fertilisation of ideas between nations and healthy competition, both from national fusion research programmes and private companies striving to be the first to reach commercial viability. Private companies have the ability to build prototypes faster and try out more unusual ideas than publicly funded projects. The latter, simply because they are funded with taxpayer money, tend to have to be more cautious in what technological approaches they take, although this is balanced with the advantage that the cash stream is steady. While being extremely prudent with the cash is a sensible policy decision, from a scientific perspective, this caution might not yield the breakthroughs needed to finally reach the goal of generating electricity from fusion. Smaller companies that are able to explore fusion quicker or in more flexible ways could end up holding the key to success. That said, neither governments nor the private sector are likely to achieve the end result completely alone.

Any fusion power plants will cost in the order of billions of pounds, so in the United Kingdom, as in the United States, it is recognised that a mixture of public and private investment, both in terms of finances and resources, will be needed. In terms of a skilled workforce alone, the numbers required will have to rapidly grow. To give an example of the size of workforce that might be needed if the United Kingdom were to move to having fusion power plants generating a meaningful proportion of its electricity, at the height of the nuclear fission industry in the 1960s there were 40,000 people employed by the UKAEA in the sector. The current UKAEA fusion energy workforce stands at around 2,000 individuals.

This clearly indicates that there is a need for a lot more fusion expertise in the United Kingdom. Recognising this, as part of the government's Fusion Foundations programme, launched in 2020, the UKAEA's apprenticeship scheme will be expanded, as will its teaching facilities. The programme aims to have 1,000 apprentices training in fusion and related technologies by 2025 at sites across the country, including the Culham Fusion Campus, which in itself is being transformed into a global fusion hub. The first major milestone in that journey was the 2021 announcement by Canadian private company General Fusion that it would build its demonstration reactor at the Culham site, as we saw in Chapter 4.

Playing by the rules

Of course, working out sufficient funding for fusion development is not the only sum that will need to add up. If we are to move to viable fusion power plants, the end product of fusion-generated electricity per kilowatt-hour needs to be affordably priced for the consumer. Otherwise, fusion energy will not be a good commercial prospect. This means that what is known as the Levelised Cost of Electricity, which is a measure of how much it costs a power plant to generate electricity over its entire lifetime, including its construction and decommissioning phases, needs to be taken into consideration.

There is also another aspect which will need to be addressed: health and safety regulation. As important as the technology is, any breakthroughs in fusion will only be able to be used if the right sort of regulation is in place to enable the safe development and – if all goes to plan – deployment of fusion power generation. Each nation wanting to play a part in the potential fusion revolution has therefore been looking at regulations both for the current wave of experiments and for future power stations should fusion prove commercially viable.

Of course, whether fusion ever proves practicable is the whole crux of the matter. So, exactly how long are we likely to have to wait to find out if fusion is the answer – or at least part of the solution – to reducing carbon emissions from electricity generation?

Are we nearly there yet?

There was a time when TV interviews of scientists talking about their latest breakthrough had an element of predictability. You just knew that when pressed on how soon we would see this research become a commercial product, or be applied on a day-to-day basis, that the scientist would answer 'five years'. Sometimes these inventions never made it out of the lab. Those that did often took much longer than predicted to become a viable product or method. Fusion energy has historically suffered a similar fate.

As mentioned earlier, it is likely to be a combination of private and public sector work that will lead to the first commercially viable fusion power plant. This is because the different approaches that each are able to bring will allow for the maximum amount of innovation. Healthy rivalry between companies and nations should drive innovation more rapidly. But just how quickly can we expect to see tangible results?

In the case of ITER, the global Covid-19 pandemic has, somewhat inevitably, introduced delays. Supply chains for components were affected as some factories ceased production and transportation was limited. This is likely to push the date for achieving the first plasma beyond the expected 2025 deadline, although the attainment of full fusion power is expected to remain on schedule for 2035.

When asked about the timescale for fusion power first making it onto a grid, seventeen of the twenty-three companies surveyed in the Global Fusion Industry in 2021 report (produced by the Fusion Industry Association and the UKAEA) felt this would occur in the 2030s. Meanwhile, eleven of the companies surveyed expected fusion to be providing space propulsion by the 2030s, while eight companies felt this would not be possible until the 2040s.

What if?

If the hopes and dreams of the scientists and engineers working on these projects come true, within a couple of decades, nuclear fusion may be playing a major role in the drive to reduce the world's carbon emissions. But only time will tell if one of these approaches will win out over another. Or if several, or indeed a combination of technologies, will ultimately prove to be the blueprint for fusion power generation.

It should be noted that fusion faces stiff competition from other green energy sources. Alongside the race to viable fusion energy there is, therefore, another competition to be won. This is to become the technology that plugs the gap in electricity supply that cannot practically be provided by a combination of wind and solar. At the time of writing, carbon capture and storage, stored green hydrogen and bioenergy with carbon capture are the main contenders in this parallel contest. Nuclear fission, which is low-carbon, will also play a role in meeting the electricity demands of some nations. In the United Kingdom, for example, where fission reactors are considered a 'clean' energy option, in January 2022, the government announced £100 million ($126 million) of funding to support the continued development of the Sizewell C nuclear power plant in Suffolk.

Before too long we will see if fusion wins the competition with its green rivals. Or whether, not least because unlike fusion these technologies have already been proven to work, fusion will be pushed into second place or even knocked off the future energy podium altogether.

Another scenario is that we learn it is impossible to build a viable fusion power plant. Although that sounds like an unthinkable outcome after all this investment of time, skills and money, from a scientific perspective, a null result can be just as important as a positive one. If that – albeit slightly nightmarish – situation does come true, it does not take away the spin-offs that have come from the journey. As we have seen, a wide range of technological developments are already being used by other industries or different sectors of the nuclear industry thanks to the pursuit of fusion power. More developments are surely still to come, and thousands of jobs have been created and supported by fusion research projects. So, in no sense will the pursuit of this single, tricky goal be completely wasted.

Whatever the eventual outcome, and whenever that might be, one thing is for sure. There is no lack of effort going in from the scientists, engineers, policymakers and funders attempting to create our very own version of the Sun here on Earth. Let's hope they succeed.

FURTHER READING

Chapter 1: What is Fusion?

Rutherford: http://rutherford.org.nz/ – a website with interesting biographical information about Ernest Rutherford, including a list of his scientific publications.

Chapter 2: Written in the Stars

Atoms for Peace Conference, 1958: https://www.ncbi.nlm.nih.gov/pmc/articles/PMC1373588/ – an overview of the published proceedings from this United Nations international conference.

Lasers Across the Cherry Orchards, Mike Forrest (self-published, 2011) – a fascinating first-hand insight into the early days of fusion research in the United Kingdom and the visit to the Tokamak T3 group in the Soviet Union.

Joint European Torus: https://ccfe.ukaea.uk/research/joint-european-torus/ – webpage giving more information about JET, including an interactive diagram showing the main reactor components.

Chapter 3: Magnetic Attraction

ITER: https://www.iter.org/ – a wealth of information about the ITER project and fusion in general.

Korea Superconducting Tokamak Advanced Research: https://www.kfe.re.kr/eng/pageView/103 – webpage about the KSTAR project at the Korea Institute of Fusion Energy, including a virtual video tour of the reactor facility.

Wendelstein 7-X stellarator: https://www.ipp.mpg.de/w7x – webpage about the stellarator at the Max Planck Institute for Plasma Physics, including a virtual tour of the facility with embedded videos by scientists describing their work.

CTFusion: https://ctfusion.net/technology/ – details about the company's approach to fusion, including graphics.

Chapter 4: Competing Technologies

National Ignition Facility: https://lasers.llnl.gov/about/what-is-nif – this section of the Lawrence Livermore National Laboratory website gives insights into the scientific and technological advances being made at the NIF facility in the United States.

General Fusion: https://generalfusion.com/ – more information about the technologies being researched by the company, including videos and graphics.

TAE Technologies: https://tae.com/ – company website including brief descriptions of their research areas and links to their published research in scientific journals.

Chapter 5: An Engineering Feat

Magnets at ITER: https://www.iter.org/mach/Magnets – this section of the ITER website gives detailed information on how their fusion reactor's magnet system will function, as well as providing updates on the construction of the system.

Beryllium: https://materion.com/products/beryllium-products – detailed information about beryllium and its applications in industry and medicine from Materion, the world's leading mining and production company for the metal.

Slowed-down video footage of the plasma inside the prototype fusion reactor at the University of Washington: https://www. washington.edu/news/2021/07/22/gaming-graphics-card-allows -faster-more-precise-control-of-fusion-energy-experiments/

Chapter 6: Tidying up After

Video about Nuclear Train Flask Collision Test – Operation Smash Hit (1984), YouTube: https://www.youtube.com/ watch?v=2jzugX2NMnk

'Processing of Used Nuclear Fuel': https://world-nuclear.org/ information-library/nuclear-fuel-cycle/fuel-recycling/ processing-of-used-nuclear-fuel.aspx – an overview of current reprocessing of nuclear fuel from fission reactors on the World Nuclear Association website.

Chapter 7: To Infinity and Beyond

Video about Boston Dynamics Spot robot, YouTube: https:// www.youtube.com/watch?v=wlkCQXHEgjA&t=3s

'The Fusion Driven Rocket project': https://www.nasa.gov/ directorates/spacetech/niac/2012_Phase_II_fusion_driven_ rocket/ – a description of NASA's past project investigating fusion propulsion.

Pulsar Fusion: https://pulsarfusion.com/ – the company website which includes details of their aims and progress, as well as video footage of an experimental firing of a prototype fusion engine for space travel.

Chapter 8: The Race is On

'The global fusion industry in 2021': https://www.fusionindustry association.org/about-fusion-industry – a downloadable survey of fusion companies by the Fusion Industry Association and the UK Atomic Energy Authority.

'Powering the Future: Fusion & Plasmas': https://usfusionand plasmas.org/ – a 2020 report from the Fusion Energy Sciences Advisory Committee in the United States. This gives a long-range plan for the country to deliver fusion energy and to advance plasma science.

'Towards Fusion Energy: the UK fusion strategy': https://www. gov.uk/government/publications/towards-fusion-energy-the-uk-fusion-strategy – a 2021 strategy document detailing the UK government's plans to enable the delivery of fusion power.

INDEX